Communications in Asteroseismology

Volume 155
October, 2008

User Manual for FAMIAS and DAS
Frequency Analysis and Mode Identification for AsteroSeismology
and
Database with time series for AsteroSeismology,

developed in the framework of the FP6 European Coordination Action HELAS

Edited by Wolfgang Zima

Austrian Academy
of Sciences Press

Vienna 2008 ÖAW

Communications in Asteroseismology

Editor-in-Chief: **Michel Breger**, michel.breger@univie.ac.at
Editorial Assistant: **Daniela Klotz**, klotz@astro.univie.ac.at
Layout & Production Manager: **Paul Beck**, paul.beck@univie.ac.at
English Language Editor: **Natalie Sas**, natalie.sas@ster.kuleuven.be

Institut für Astronomie der Universität Wien
Türkenschanzstraße 17, A - 1180 Wien, Austria
http://www.univie.ac.at/tops/CoAst/
Comm.Astro@univie.ac.at

Cover Illustration: Selected screenshots of the program FAMIAS

British Library Cataloguing in Publication data.
A Catalogue record for this book is available from the British Library.

Austrian Academy of Sciences Press
A-1011 Wien, Postfach 471, Postgasse 7/4
Tel. +43-1-515 81/DW 3402-3406, +43-1-512 9050
Fax +43-1-515 81/DW 3400
http://verlag.oeaw.ac.at, e-mail: verlag@oeaw.ac.at

Contents

Preface   6

DAS User Manual   7

1. HELAS Database for AsteroSeismology   9
   1.1 Introduction   9
   1.2 Features   9
   1.3 Archive data   10
   1.4 Referencing   10
   1.5 Platform   11
   1.6 Call for Contributions   11

Acknowledgements   12

FAMIAS User Manual   17

1. Introduction   19
   1.1 Overview   19
   1.2 What Data Can Be Used?   22
   1.3 Requirements   22

2. The Main Window   23
   2.1 The File Menu   23
   2.2 The Edit Menu   26
   2.3 The Tools Menu   26
   2.4 The Help Menu   26

3. The Plot Window   28

4. The Spectroscopy Modules   29
   4.1 Data Manager   29
        4.1.1 Data Sets Box   30
        4.1.2 Time Series Box   30
        4.1.3 Spectrum Box   37
        4.1.4 Plot Window   38

4.2   Fourier Analysis 39
      4.2.1   Settings Box 40
      4.2.2   List of Calculations 43
      4.2.3   Fourier Spectrum Plot 43
4.3   Least-Squares Fitting 44
      4.3.1   Settings 44
      4.3.2   List of Frequencies 47
      4.3.3   Least-Squares Fit plot 49
4.4   Line Profile Synthesis 50
      4.4.1   Theoretical background 50
      4.4.2   Stellar Parameters 53
      4.4.3   Line Profile Parameters 55
      4.4.4   Pulsation Mode Parameters 56
      4.4.5   General Settings 56
4.5   Mode Identification 58
      4.5.1   The FPF method 59
      4.5.2   Optimisation settings for the FPF method 60
      4.5.3   Practical information for applying the FPF method . . . 62
      4.5.4   The moment method 63
      4.5.5   Practical information for applying the moment method . 64
      4.5.6   Setting of parameters 65
      4.5.7   Stellar Parameters 66
      4.5.8   Pulsation Mode Parameters 67
      4.5.9   Line Profile Parameters 68
      4.5.10 Optimisation Settings 69
      4.5.11 General Settings 71
4.6   Results 73
      4.6.1   Best models 74
      4.6.2   Chi-square plots 74
      4.6.3   Comparison between fit and observation 74
      4.6.4   List of calculations 75
4.7   Logbook 75
4.8   Tutorial: Spectroscopic mode identification 76
      4.8.1   Import spectra 76
      4.8.2   Select dispersion range 77
      4.8.3   Convert from wavelength to Doppler velocity 77
      4.8.4   Compute signal-to-noise ratio and weights 78
      4.8.5   Compute moments 78
      4.8.6   Interpolate on common dispersion scale 79
      4.8.7   Compute line statistics 79
      4.8.8   Searching for periodicities 79

4.8.9   Mode identification 85

5. The Photometry Modules   93
   5.1 Data Manager 93
        5.1.1 Data Sets Box 93
        5.1.2 Time Series Box 94
        5.1.3 Plot window 95
   5.2 Fourier Analysis 96
        5.2.1 Settings Box 96
        5.2.2 List of Calculations 97
        5.2.3 Fourier Spectrum Plot 98
   5.3 Least-Squares Fitting 99
        5.3.1 Settings 99
        5.3.2 List of Frequencies 101
   5.4 Mode Identification 102
        5.4.1 Theoretical background 102
        5.4.2 Approach for mode identification in FAMIAS 104
        5.4.3 Observed values 106
        5.4.4 Stellar model parameters 107
   5.5 Results 109
        5.5.1 List of Calculations 109
        5.5.2 Settings 109
        5.5.3 Mode Identification Report 110
        5.5.4 Mode Identification Plots 110
   5.6 Logbook 111
   5.7 Tutorial: Photometric mode identification 112
        5.7.1 Importing and preparing data 112
        5.7.2 Searching for periodicities 113
        5.7.3 Mode identification 115

Bibliography   118

Acknowledgements   119

Comm. in Asteroseismology
Vol. 155, 2008

Preface

At the time of the definition of the deliverables for the HELAS network activity NA5: Asteroseismology, a consultation was done within the European astero-seismology community during CoRoT Week 9 in December 2005 at ESA/ESTEC. The majority of asteroseismologists present at that meeting were in favour of developing a software tool for the identification of kappa-driven oscillation modes in main-sequence stars, from multi-colour photometry and/or high-resolution spectroscopy. It was felt that the need for such a type of tool was much greater than for any other one, given that such a package is not available while various frequency analysis and modelling tools were already offered. Ideally, a database of time series for mode identification would come along with such a package, so that newcomers in the field of asteroseismology (at Master, PhD or even postdoc level) as well as lecturers would have a complete toolkit for mode identification at their disposal.

The current special volume of *Communications in Asteroseismology* provides the user manuals of both released tools. We present the manuals of the *Database for AsteroSeismology* (DAS) which was defined and implemented by Dr. Roy Østensen and of the *Frequency Analysis and Mode Identification for AsteroSeismology* (FAMIAS) developed by Dr. Wolfgang Zima, both at the Institute of Astronomy of Leuven University, which is the lead institute of the HELAS Workpackage NA5. Both authors have committed to maintain and update DAS and FAMIAS for the whole duration of HELAS.

We refer to the NA5 website[1] for additional HELAS NA5 asteroseismology deliverables prepared by the Porto and Wrocław teams. It concerns grids of non-adiabatic observables, atmospheric model parameters, grids of stellar models and isochrones as well as their frequencies of oscillation, and, finally, model comparison tools and documentation.

We hope that this ensemble of asteroseismology tools is of use for the community.

Conny Aerts, Chairwoman of NA5, Leuven, 15 August 2008.

---

[1] http://www.ster.kuleuven.be/~zima/helasna5/

Comm. in Asteroseismology
Vol. 155, 2008

DAS User Manual

Roy Østensen
Instituut voor Sterrenkunde, K.U. Leuven
B-3001 Leuven, Belgium
e-mail: roy@ster.kuleuven.be
http://newton.ster.kuleuven.be/~roy/helas/

1. HELAS Database for AsteroSeismology

1.1 Introduction

The HELAS Database for AsteroSeismology (DAS) is one of the deliverables of
the Work Package NA5: Asteroseismology of the European Coordination Action in Helio and Asteroseismology (HELAS[1]). The DAS aims to provide easy
access to publicly available asteroseismological timeseries data, both photometric and spectroscopic. In particular, the DAS and the HELAS software package
FAMIAS are ideally suited to train Master and PhD students in asteroseismic
data analysis and to build longterm datasets. The number of stars in the system
is still limited and reflects the willingness of data owners to provide their data
after publication. Work continues to populate the database with contributions
from the community, and at present the number of stars in the database is 82
(Tables 1 and 2).

1.2 Features

Before getting access to the database, the user must agree to the conditions
of use, which obliges the user to refer to a source publication provided with
each dataset, whenever archive data is used in an article. The database search
interface (Figure 1) includes search by variable class, name or coordinates.
Output tables (Figure 2) are generated in HTML with links to automatically
generated finding charts, the Aladin viewer, and a detailed data sheet (Figure 3)
that displays catalogue data for each target, together with a DSS image of the
source. All stars have been added with a number of identifiers including the
common or constellation name, HD catalog number, Hipparcos number, BD
catalog name and others, making it easy to find a particular star in the database.
The database currently recognises 15 different classes of variable stars, but not
all classes have any entries yet. Table 3 provides the keys to the variable star
classes used in the database (and in Tables 1 and 2), and also summarises the
total number of stars in each class.

---

[1] http://www.helas-eu.org

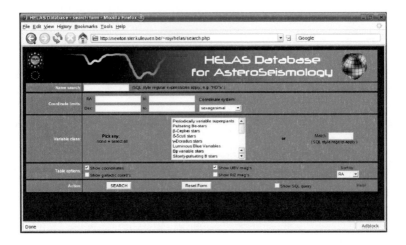

Figure 1: The database search interface.

1.3   Archive data

At the bottom of the data sheet, a table is generated with all datasets for this star in the database (see Figure 3 for an example). Each dataset is associated with a type identifier, usually 'spectra' or 'rv' to distinguish between sets that contain actual spectroscopy (usually a small section of a high resolution spectrum or several lines combined into one) or radial velocities derived from spectroscopy. Photometric datasets have also been included, and since they can be very different in nature, they have been given different identifiers such as 'most', 'ultracam', 'geneva', 'wet' and so on. In Table 1, several stars with both spectroscopic and photometric data are listed.

1.4   Referencing

With each dataset entry, there is an associated README file, which, in addition to describing the format of the data provided, gives one or more references to articles that use and describe the dataset. This reference, or the most important one if there are several, is also provided in the data table as a link that will lead to the relevant paper. The final entry in the dataset table is a link to the actual dataset, normally a compressed TAR-file. The dataset can contain either tables of timeseries data or in the case of spectra, individual files for each measurement. The README-file is always included in the archive file.

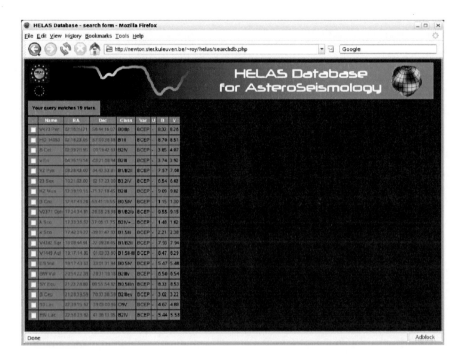

Figure 2: The result of a search for β Cephei stars in the database.

1.5   Platform

The database runs under MySQL (www.mysql.com; an open source database) with interfaces in Perl for uploading the database entries. The WWW inter-face uses HTML forms and tables generated by PHP. The DAS is hosted at http://newton.ster.kuleuven.be/~roy/helas/ and also available from the HELAS platform[2] through the NA5 website link.

1.6   Call for Contributions

Anybody who wishes to contribute published data, spectroscopic or photomet-ric, on any particular star, is kindly asked to contact the author by e-mail to roy@ster.kuleuven.be.

---

[2]http://www.helas-eu.org

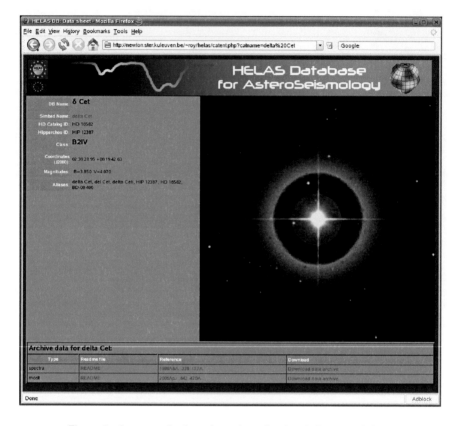

Figure 3: An example data sheet, here for the β Cep star δ Ceti.

Acknowledgements

RØ has been supported by the FP6 European Coordination Action HELAS and by the Research Council of the University of Leuven under grant GOA/2003/04.

Table 1: Variable stars with data in the DAS.

DB name	HD number	Class	Data type
HD 215	HD 215	GDOR	geneva
HD 277	HD 277	GDOR	geneva
EK Psc		SDBV	ultracam
V746 Cas	HD 1976	SPB	geneva
HD 2842	HD 2842	GDOR	geneva
53 Psc	HD 3379	SPB	geneva
PG 0101+039		SDBV	most
HD 7169	HD 7169	GDOR	geneva
FO Cet	HD 12901	GDOR	spectra, geneva
V354 Per	HD 13745	SPB	geneva
V473 Per	HD 13831	BCEP	geneva
HD 14053	HD 14053	BCEP	geneva
delta Cet	HD 16582	BCEP	spectra, most
53 Ari	HD 19374	SPB	geneva
V576 Per	HD 21071	SPB	geneva
IP Per	HD 278937	DSCUT	multisite
HD 23874	HD 23874	GDOR	geneva
tau08 Eri	HD 24587	SPB	spectra
DO Eri	HD 24712	WR	wet
V1133 Tau	HD 25558	SPB	geneva
GU Eri	HD 26326	SPB	spectra
V1143 Tau	HD 28114	SPB	geneva
V1144 Tau	HD 28475	SPB	geneva
nu Eri	HD 29248	BCEP	spectra
V350 CMa	HD 48501	GDOR	spectra, geneva
V450 Car	HD 53921	SPB	spectra
MM CMa	HD 55522	BPV	spectra
DO Lyn	HD 62454	GDOR	geneva
rho Pup	HD 67523	DSCUT	spectra
NO Vel	HD 69144	SPB	spectra
EF Lyn	HD 69715	GDOR	geneva
YZ Pyx	HD 71913	BCEP	geneva
HY Vel	HD 74560	SPB	spectra
omicron Vel	HD 74195	SPB	spectra
HD 74504	HD 74504	GDOR	geneva
V335 Vel	HD 85953	SPB	spectra
HD 86358	HD 86358	GDOR	geneva
23 Sex	HD 89688	BCEP	geneva
V514 Car	HD 92287	SPB	spectra
xi Hya	HD 100407	SLR	rv
V863 Cen	HD 105382	BPV	spectra

Table 2: Variable stars (cont'd.)

DB name	HD number	Class	Data type
HD 105458	HD 105458	GDOR	geneva
FG Vir	HD 106384	DSCUT	spectra
DD CVn	HD 108100	GDOR	geneva
KZ Mus	HD 109885	BCEP	geneva
beta Cru	HD 111123	BCEP	spectra
MP Com	HD 113867	GDOR	geneva
V869 Cen	HD 123515	SPB	spectra
alfa Cen B	HD 128621	SLR	rv
V1019 Cen	HD 131120	BPV	spectra
FK Boo	HD 138003	SPB	geneva
IU Lib	HD 138764	SPB	spectra
d Lup	HD 138769	BPV	spectra
PT Ser	HD 140873	SPB	spectra
epsilon Oph	HD 146791	SLR	rv
zeta Oph	HD 149757	BE	most
J1717+5805		SDBV	ultracam
V2371 Oph	HD 157485	BCEP	geneva
lambda Sco	HD 158926	BCEP	spectra
kappa Sco	HD 160578	BCEP	spectra
HD 163830	HD 163830	SPB	most
HD 163899	HD 163899	PVSG	most
V3984 Sgr	HD 163868	BE	most
V4382 Sgr	HD 165812	BCEP	geneva
V1402 Aql	HD 177230	WR	most
V4198 Sgr	HD 177863	SPB	spectra
V338 Sge	HD 169820	SPB	geneva
V1449 Aql	HD 180642	BCEP	geneva
2 Vul	HD 179588	SPB	geneva
ES Vul	HD 180968	BCEP	geneva
V4199 Sgr	HD 181558	SPB	spectra
V377 Vul	HD 182255	SPB	geneva
V1473 Aql	HD 191295	SPB	geneva
BW Vul	HD 199140	BCEP	spectra
SY Equ	HD 203664	BCEP	geneva
beta Cep	HD 205021	BCEP	spectra
HD 206540	HD 206540	SPB	geneva
16 Peg	HD 208057	SPB	geneva
10 Lac	HD 214680	BCEP	geneva
xi Oct	HD 215573	SPB	spectra
EN Lac	HD 216916	BCEP	spectra
V394 And	HD 222555	SPB	geneva

Table 3: Variable star classes used in the DAS.

Class	Reference	Total
PVSG	Periodically variable supergiants	1
BE	Pulsating Be-stars	2
BCEP	β-Cephei stars	19
DSCUT	δ-Scuti stars	3
GDOR	γ-Doradus stars	14
LBV	Luminous Blue Variables	0
BPV	Bp variable stars	4
SPB	Slowly-pulsating B stars	30
SLR	Solar-like oscillations in red giants	3
SDBV	Pulsating subdwarf B stars	3
DAV	Pulsating DA white dwarfs	0
GWVIR	GW-Virginis stars	0
ROAP	Rapidly oscillating Ap stars	0
WR	Wolf-Rayet stars	2
CV	Cataclysmic variables	0

FAMIAS User Manual

Wolfgang Zima
Instituut voor Sterrenkunde, K.U. Leuven
B-3001 Leuven, Belgium
e-mail: zima@ster.kuleuven.be
http://www.ster.kuleuven.be/~zima/famias

1. Introduction

FAMIAS (Frequency Analysis and Mode Identification for AsteroSeismology) is a collection of state-of-the-art software tools for the analysis of photometric and spectroscopic time series data. It is one of the deliverables of the Work Package NA5: Asteroseismology of the European Coordination Action in Helio- and Asteroseismology (HELAS).

Two main sets of tools are incorporated in FAMIAS. The first set allows to search for periodicities in the data using Fourier and non-linear least-squares fitting algorithms. The other set allows to carry out a mode identification for the detected pulsation frequencies to determine their pulsational quantum numbers, the harmonic degree, ℓ, and the azimuthal order, m. The types of stars to which FAMIAS is applicable are main-sequence pulsators hotter than the Sun. This includes the Gamma Dor stars, Delta Sct stars, the slowly pulsating B stars and the Beta Cep stars - basically all pulsating main-sequence stars, for which empirical mode identification is required to successfully carry out asteroseismology.

This user manual describes how to use the different features of FAMIAS and provides two tutorials that demonstrate the usage of FAMIAS for spectroscopic and photometric mode identification.

1.1 Overview

The following key features are provided by FAMIAS:

- Search for periodicities in photometric/spectroscopic time series using Fourier analysis and multi-periodic least-squares fitting techniques.

- Spectroscopic mode identification using the moment method (Briquet & Aerts 2003) and Fourier parameter fit method (Zima 2006)

- Photometric mode identification using the method of amplitude ratios and phase differences based on pre-computed model grids (Balona & Stobie 1979; Watson 1988; Cugier et al. 1994; Daszyńska-Daszkiewicz et al. 2002).

- Efficient usage of multi-core processors with parallel computing.

The user interface of FAMIAS is structured into different *tabs* that contain the modules dedicated to the different tools. The tabs can be selected by clicking on their descriptive name. The two main modules are for the spectroscopic and photometric analysis. Each of these modules is subdivided into tools for data management, frequency searching and mode identification.

In the following, the different available tabs are briefly described.

Spectroscopy Tabs

- **Data Manager**
  Edit the time series of spectra or moments, perform statistics, compute line moments, examine the spectra, extract spectral lines, etc.

- **Fourier Analysis**
  Compute a Fourier analysis (Discrete Fourier Transformation) for each pixel of a spectrum (pixel-by-pixel) or for the different line moments in order to detect periodicities and their statistical significance.

- **Least-Squares Fitting**
  Compute a non-linear multi-periodic least-squares fit across a line profile (pixel-by-pixel) or for the different line moments and pre-whiten the data.

- **Line Profile Synthesis**
  Compute a time series of theoretical line profiles of a radially or non-radially pulsating star.

- **Mode Identification**
  Identify pulsation modes by means of the Fourier parameter fit method or the moment method.

- **Results**
  The results of the mode identifications are displayed and logged in this tab.

- **Logbook**
  Log of all actions that were carried out in the spectroscopy module.

Photometry Tabs

- **Data Manager**
  Edit and modify the photometric time series.

- **Fourier Analysis**
  Compute a Fourier analysis (DFT) and determine the statistical significance of detected frequency peaks.

- **Least-Squares Fitting**
  Compute a non-linear multi-periodic least-squares fit and pre-whiten the data.

- **Mode Identification**
  Carry out a photometric mode identification with the method of amplitude ratios and phase differences in different photometric passbands.

- **Results**
  The results of the mode identification are displayed and logged in this tab.

- **Logbook**
  Log of all actions of the photometry module.

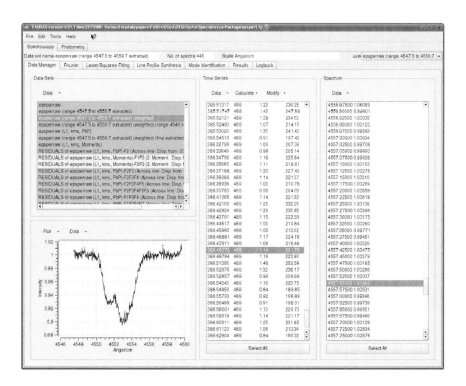

Figure 1: Screenshot of FAMIAS.

1.2   What Data Can Be Used?

The spectroscopic as well as the photometric data that can be analysed with FAMIAS must fulfill specific quality criteria and must have been fully reduced. More specifically, this implies the following requirements:

- **Requirements for spectroscopic data**

   - Time series of fully reduced and normalised spectra, including barycentric time and velocity correction
   - Dispersion better than 40000
   - Signal-to-noise ratio higher than 200
   - Unblended absorption line

- **Requirements for photometric data**

   - Time series of fully reduced differential photometric data, including barycentric time correction
   - Multi-colour data in Strömgren, Johnson/Cousins, or Geneva filters for mode identification
   - Milli-mag precision

1.3   Requirements

FAMIAS has been written in the programming language C++. For the graphical user interface, the open source version of the Qt 4 library (from Trolltech[1]) has been adopted. This combination enabled the development of a software tool that requires high computational speed in combination with the ability to create cross platform versions for Linux and Mac OS X. FAMIAS also features a built-in help system with an extended manual describing the tools and providing introductory tutorials.

The homepage of FAMIAS[2] provides the possibility to download the software, read the on-line documentation, and to submit bug reports.

---

[1] http://www.trolltech.com
[2] http://www.ster.kuleuven.be/∼zima/famias

2. The Main Window

After the start-up of FAMIAS, the *Data Manager Tab* of the *Spectroscopy Module* is shown. If you wish to work with photometric data, you have to switch to the *Photometry Module*. In this chapter, the main menu entries of FAMIAS are described.

2.1 The File Menu

This menu contains entries for opening and saving project files and importing time series of spectra or photometric measurements. A session with FAMIAS can be saved as a project file. Such a project file contains all the data included in the current session of FAMIAS. The following entries are available in this menu.

- **New Project**
  Creates a new, empty project. All entries of the current FAMIAS session will be cleared.

- **Open Project**
  Opens an existing project. All current entries in FAMIAS will be cleared and replaced by the opened project.

- **Recent Projects**
  Shows a list of previously opened project files.

- **Save Project**
  Saves the current session of FAMIAS as a project file with the current file name (if existent). In a project the complete content of all modules of FAMIAS is saved. This includes time series data, diagrams, results from the analyses and the logbook.

- **Save Project as**
  Saves the current session of FAMIAS as a project file with a new file name.

- **Import Set of Spectra**
  Opens a dialogue to import a time series of spectra. The selected file must be in ASCII format and list the filenames of the spectra, the observation

times (Heliocentric Julian Date), and optionally the weights of the spectra and their signal-to-noise ratio. The separate spectrum files must also have ASCII format and require the following two column structure: wavelength or Doppler velocity in km s^{-1} and normalised intensity. File headers can be skipped during the import of the files.

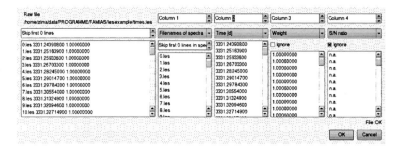

Figure 2: Screenshot of the dialogue for importing a time series of spectra.

Figure 2 shows a screenshot of the dialogue for importing spectra. The left column shows the raw data file. You can indicate the number of header lines to be skipped (*Skip first X lines*). At the top of the right part of the window you can select the column number which contains the data type selected in the box below. If your spectrum files have headers, you can skip them by choosing the number of lines to skip. The time must be in units of days (Heliocentric Julian Date), and the weights are point weights per spectrum. Click on *Ignore* if no column with weights is available. All weights are then automatically set to 1. Optionally, a column with the signal-to-noise ratio per spectrum can be imported. These values are used to estimate the uncertainties of the line moments in FAMIAS. The SNR can also be estimated within FAMIAS (see p. 32).

If your selections are valid and the file structure is acceptable for importing, the text on the lower right will read *File OK*. Otherwise, you have to check the structure of your data file. Still, if there is a problem with the structure of the spectrum files, FAMIAS might give an error message during importing, indicating in which file the read-in error occurred.

After you click on *OK*, you must select the dispersion scale of your spectra (units of Ångstrom or Doppler velocity (km s^{-1})). If the import was successful, you will see the imported data in the *Data Manager Tab* of the spectroscopy module.

- **Import Light Curve(s)**
  Opens a dialogue to import a time series of photometric data. You can import several files simultaneously by making a multi-selection in the file import window (by pressing the Shift or Ctrl-key when selecting files). Imported files are required to be in ASCII format and must consist of two or three columns separated by spaces or tabulators. The following two columns are mandatory: time in days (Heliocentric Julian Date) and differential magnitude. A third column can consist of weights for the single measurements. A file header can be skipped during import.

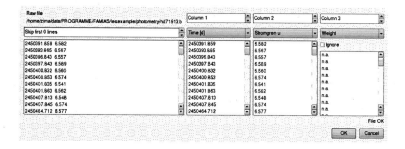

Figure 3: Screenshot of the dialogue for importing light curves.

Figure 3 shows a screenshot of the dialogue for importing light curves. The left column shows the raw data file. You can indicate the number of header lines to be skipped (*Skip first X lines*). At the top of the right part of the window you can select the column number which contains the data type selected in the box below. The time must be in units of days. You must select a passband for the magnitude, and the weights are point weights per data point. Click Ignore if no column with weights is available. All weights are then automatically set to a value of 1.

If your selections are valid and the file structure is acceptable for importing, the text box on the lower right will display *File OK*. Otherwise, you will have to check the structure of your data file.

If the import was successful, you will see the imported data in the *Data Manager* of the photometry module.

- **Quit**
  Exit FAMIAS.

2.2 The Edit Menu

The items in this menu provide the possibility to clear the input fields, plots, and stored data of selected tabs of FAMIAS.

- **Clear Spectroscopy Tabs**
  Clear all data in the selected tabs of the spectroscopy module.

- **Clear Photometry Tabs**
  Clear all data in the selected tabs of the photometry module.

2.3 The Tools Menu

This menu provides some useful tools that are related to asteroseismology and mode identification.

- **Stellar Rotation**
  Compute the equatorial rotational velocity, rotation period or rotation frequency, theoretical critical Keplerian break-up velocity or critical $v \sin i$, and the critical minimum inclination for a given stellar mass, radius, $v \sin i$, and inclination angle.

- **Pulsation Parameters**
  Compute the horizontal-to-vertical amplitude ratio, the frequency in the stellar frame of reference, and the rotation frequency and the ratio of the rotation to the pulsation frequency for a given pulsation mode, stellar mass, radius, $v \sin i$, and inclination angle.

2.4 The Help Menu

The *Help Menu* provides access to the FAMIAS-manual, enables to submit bug reports and provides general information about the software.

- FAMIAS **Help**
  This opens the built-in user manual of FAMIAS. The manual is regularly updated with new versions of FAMIAS.

- **Update Information**
  This shows a list containing update information about the current and previous versions of FAMIAS.

- **Report a Bug**
  Provides a link to the webpage of FAMIAS, where bug reports can be submitted on-line.

- **Copyright and User Agreement**
  View general copyright information for FAMIAS and the user agreement.

- **About FAMIAS**
  Provides some general information about FAMIAS.

- **About Qt**
  Provides an information box about the version of Qt that was used for the current version of FAMIAS. The graphical user interface of FAMIAS has been programmed with the Trolltech Qt-library.

3. The Plot Window

A plot can be zoomed in by pressing the left mouse button while moving the mouse to draw a zoom box. Pressing the right mouse button zooms out. Keep the middle mouse button pressed to pan the plot.
The following commands are available in the menu Plot:

- **Refresh Plot/Show All**
  Refresh the contents of the current plot.

- **Set Viewport** Set the viewport of the current plot.

- **Detach Plot**
  Open current plot in a new window.

- **Print Plot**
  Print the current plot.

- **Export Plot To PDF**
  Write the current plot into a PDF file. If this is a multi-plot (e.g., zero-point, amplitude and phase from least-squares fitting), the sub-plots will be written into separate files.

The following commands are available in the menu Data:

- **Overplot**
  If this option is checked, the plot window is not cleared when a new plot is drawn.

- **Show Original and Fit**
  If a least-squares fit has been computed for these data, this option shows the original data (spectrum, line moments or light curve) and the multi-periodic least-squares fit.

- **Show Residuals**
  Only available if the current data set consists of residuals (pre-whitened in the *Least-Squares Fitting Tab*). The original data minus the least-squares fit are shown.

- **Show Phase Plot**
  Plot the data phased with the indicated frequency.

4. The Spectroscopy Modules

After the start-up of FAMIAS, the *Data Manager Tab* of the *Spectroscopy Module* is shown. The *Spectroscopy Module* contains the tools that are required to search for frequencies in time series of spectra and to carry out a spectroscopic mode identification. Additionally, synthetic line profile variations of a multi-periodic pulsating star can be computed. The tools are located in tabs that have the following denominations: *Data Manager, Fourier, Least-Squares Fitting, Line Profile Synthesis, Mode Identification, Results,* and *Logbook.* These tools are described in the following sections.

4.1 Data Manager

The *Data Manager Tab* provides information about the data that have been imported and permits to edit the data, calculate statistics, compute moments of a spectral line, set the weights of individual spectra, or extract a line using sigma-clipping. The window is structured into three data boxes and one plot window. A menu is located above each box. In the *Data Sets Box* you can select the time series of spectra you want to work with. The *Time Series Box* shows the time, number of dispersion bins, weight, and optionally the signal-to-noise ratio of all spectra of the selected data set. The *Spectrum Box* shows the dispersion and intensity of the spectrum currently selected in the *Time Series Box.* The *Plot Window* shows the currently selected spectrum, statistics of a spectrum (mean or standard deviation), or a time series of moments (if selected in the *Data Sets Box*). A screenshot of the *Data Manager Tab* is shown in Figure 1.

Once you have successfully imported a set of spectra, its name will be added to the list of data sets (*Data Sets Box*). The times of measurements, number of wavelength bins, and the weight of each spectrum will be listed in the *Time Series Box.* The *Plot Window* will remain empty until you click on one of the spectra in the *Time Series Box.* In this case, the dispersion (in Ångstrom or km s^{-1}, dependent on your selection) and intensity of the selected spectrum will be listed in the *Spectrum Box* and the spectrum will be plotted as a blue line in the *Plot Window.* You can select multiple wavelength bins in the *Spectrum Box.* They will be displayed as red crosses in the *Plot Window.*

4.1.1   Data Sets Box

This box shows a list of the different data sets that have been imported or created. The data can consist of a time series of spectra (green background) or of line moments (yellow background). To select a data set, click on it or select it in the combo box at the top right of the information bar. The selected data set is used for all operations of FAMIAS. The following commands can be selected in the *Data Menu*:

- **Remove Data Set**
  Removes the currently selected data set from the list.

- **Rename Data Set**
  Renames the currently selected data set.

- **Export Data Set**
  The currently selected data set will be exported as ASCII-file(s) to the disk. The suffix of the files has to be entered by the user. For a time series of spectra the exported files will have the following structure: One file, called times.suffix consisting of a list of three columns, namely spectra filenames, times and weights. Each spectrum of the time series will be written into a separate ASCII file and called number.suffix, where number is a running counter. If a data set of line moments is exported, a single ASCII file having the following four columns is created: time, moment value, uncertainty, and weight.

- **Combine Data Sets**
  Combines the selected data sets to a new single time series. The data sets to be combined must have the same units of dispersion. Moreover, all times of measurement have to differ.

- **Change Dispersion Scale**
  Select wavelength in Ångstrom or Doppler velocity in km s^{-1} as dispersion scale of the current data set of spectra. A conversion between the two scales can be carried out in the *Modify Menu* of the *Time Series Box*.

4.1.2   Time Series Box

The content of this list depends on the selected data set. If a time series of spectra is selected in the *Data Sets Box*, the list will consist of three (or four) columns: times of measurement, number of wavelength bins, and weight (and optionally signal-to-noise ratio). If a moment time series is selected in the *Data Sets Box*, the list will consist of times of measurement, moment value, and weight.

Click on an item of the list to display dispersion and intensity of the selected spectrum in the *Spectrum Box*. The selected spectrum will also be displayed in the *Plot Window*. Multiple spectra can be selected by clicking with the left mouse button on several items in the list while pressing the Ctrl-key or the Shift-key. All items can be selected by pressing *Select All*. *Only items that have been selected in this list (with blue background) are taken into account for the data analysis (e.g., Fourier analysis or least-squares fitting).*

The following commands are available in the *Data Menu*:

- **Edit Data**
  Opens a table of times and weights in a new window with the possibility to edit these values. Modifications can be written to the current data set.

- **Copy Selection to New Set**
  A new data set with currently selected spectra is created and written to the *Data Sets Box*. Use this option to create subsets of your data.

- **Remove Selection**
  The currently selected spectra are removed from the time series/data set.

- **Extract Dispersion Range**
  A new data set with the currently selected spectra and the indicated dispersion range is created and written to the *Data Sets Box*. Use this option, e.g., to cut out certain spectral lines from your data set.

The following commands are available in the *Calculate Menu*:

- **Mean Spectrum**
  The weighted temporal mean for each pixel of the selected spectra is computed and displayed in the *Plot Window*. Important: All spectra must have the same dispersion scale, i.e., they must be interpolated on a common scale (use the tool *Interpolate Dispersion* in the *Modify Menu*).

- **Median Spectrum**
  The weighted temporal median for each pixel of the selected spectra is computed and displayed in the *Plot Window*. Important: All spectra must have the same dispersion scale, i.e., must be interpolated on a common scale (use the tool *Interpolate Dispersion* in the *Modify Menu*).

- **Std. Deviation Spectrum**
  The weighted temporal standard deviation (σ) for each pixel of the selected spectra is computed and displayed in the *Plot Window*. Important:

all spectra must have the same dispersion scale, i.e., must be interpolated on a common scale (use the tool *Interpolate Dispersion* in the *Modify Menu*).

- **Compute Signal-to-Noise Ratio**

  Opens a dialogue for computing the signal-to-noise ratio (SNR) of the selected spectra by making use of sigma-clipping to determine the continuum range. The calculated SNR of each spectrum can be used to set the weights of the data. In order to compute the SNR of the spectra, a sufficiently large range of continuum must be present in the selected spectra. Description of the SNR-dialogue (see Figure 4): Initially, the *Time and SNR Box* displays only the list of times of the selected spectra. Once a SNR computation has been performed, it also displays the SNR of each spectrum. Clicking on a time will show the according spectrum in the *Current Spectrum Box*.

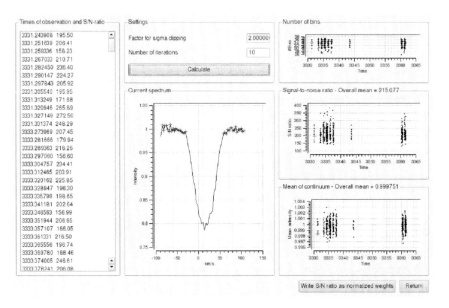

Figure 4: Screenshot of the dialogue for computing the SNR of the spectra.

In the *Settings Box*, the sigma-clipping factor and the number of iterations can be indicated and must be adapted for different data sets. Sigma clipping iteratively removes outliers of a Gaussian distribution. In this case, the sigma clipping algorithm tries to find the continuum and to exclude the spectral lines.

After clicking on *Calculate*, the position of the pixels detected as continuum is marked as red crosses in the *Current Spectrum Box*. The three plots at the right side show information about all spectra of the time series in order to check the overall results. The top plot shows the number of bins detected as continuum of each spectrum. The middle plot shows the SNR of all spectra and the overall mean SNR. The bottom plot shows the mean intensity value of the pixels detected as continuum. The latter values should be around 1. Outliers in this plot can indicate that the continuum has not been detected properly in some spectra. In this case, the settings for the sigma-clipping must be adapted.

If *Write SNR as normalised weights* is clicked, the time series is written into a new data set having normalised weights $W = (\text{SNR})^2$. Also, each spectrum of the time series is assigned its SNR-value (fourth column in the *Time Series Box*).

- **Compute Weights from SNR**
  This function computes the weights of each spectrum according to its SNR. The values of the SNR must already have been imported together with the spectra. The weights are calculated from $(\text{SNR})^2$ and normalised such that the mean value is 1.

- **Compute Moments**
  Opens a dialogue for computing line moments of the selected spectra. The dispersion scale of the spectra must be in Doppler velocity expressed in km s^{-1}. The moments time series can either be written into a new data set to the *Data Sets Box* or directly to the disk as ASCII files (if the option *Write Moments 0 to 6 in a file* has been checked).

  The nth normalised moment $<v^n>$ of a line profile $I(v, t)$, corrected for the velocity of the star with respect to the sun, at the time t is defined by

  $$<v^n>_I(t) = \frac{\int_{-\infty}^{\infty} v^n \, I(v, t) \, dv}{\int_{-\infty}^{\infty} I(v, t) \, dv}, \qquad (1)$$

  where v denotes the line-of-sight Doppler velocity of a point on the stellar surface and the denominator of this expression is equal to the equivalent width of the line. The 1st moment is the radial velocity placed at average zero, the 2nd moment describes the line variance, and the 3rd moment describes the skewness of the profile. If the shape of the line profile changes periodically due to stellar pulsations, the line moments also vary with the period of pulsation (or a sub-multiple thereof). For more details, we refer to Aerts et al. (1992).

FAMIAS can compute the uncertainties of the different line moments if the SNR of the spectra is known. The uncertainty is used to derive the χ^2-value of the theoretical moments when applying the moment method and consequently, to determine the statistical significance of obtained solutions of the mode identification. We provide here the formalism to calculate the uncertainties of the moments.

The formal uncertainty of each wavelength bin of a line profile $\sigma_{I(v,t)}$ can be derived from the signal-to-noise ratio SNR of the spectrum by

$$\sigma_{I(v,t)} = \frac{\text{SNR}}{\sqrt{I(v,t)}}. \tag{2}$$

If

$$\left[\Delta{<}v^0{>}(t)\right]^2 = \int_{-\infty}^{\infty} \left[\sigma_{I(v,t)} \; dv\right]^2 \tag{3}$$

and

$$\left[\Delta{<}v^n{>}(t)\right]^2 = \int_{-\infty}^{\infty} \left[v^n \sigma_{I(v,t)} \; dv\right]^2 \tag{4}$$

then the variance σ^2 of the moment ${<}v^n{>}$ is

$$\sigma^2_{{<}v^n{>}} = \left(\frac{\Delta{<}v^n{>}(t)}{\int_{-\infty}^{\infty} I(v,t) \, dv}\right)^2 + \left(\frac{\left[\Delta{<}v^0{>}(t)\right] \int_{-\infty}^{\infty} v^n \, I(v,t) \, dv}{\left[\int_{-\infty}^{\infty} I(v,t) \, dv\right]^2}\right)^2. \tag{5}$$

Description of the line moments dialogue (see Figure 5):

Select the dispersion range for the computation of the line moments. The range must be large enough to include the complete line profile, i.e., from continuum to continuum. Optionally, the complete dispersion range can be selected for the computations by checking *Complete range*.

The mean SNR of all spectra or the individual SNR of each spectrum is required to compute the statistical uncertainties of the moments. The mean SNR of all selected spectra at the continuum can be estimated by computing the standard deviation spectrum and taking the inverse of the standard deviation at the position of the continuum. When selecting Individual SNR, each spectrum must be assigned a specific SNR (column 4 in the *Time Series Box*).

The following procedure is strongly recommended for the calculation of line moments: compute the SNR of each spectrum with the function *Compute signal-to-noise ratio* in the *Calculate Menu*. Then extract the line profile with the function *Extract line* in the *Modify Menu*. Use the resulting spectra for computing the moments by selecting the complete

The Spectroscopy Modules 35

Figure 5: Screenshot of the dialogue for computing the line moments.

dispersion range and the option *Individual signal-to-noise ratio*. Note that the function *Extract line* determines integration boundaries for the moments that are different from one spectral line to another in order to avoid the noisy continuum in the moment computations. The use of this function is thus indispensable when the line profile is moving a lot in time.

If the user wants the nth moment to be written to the *Data Sets Box*, the moment index n must be indicated in the combo box below. The moments 0 to 6 can be exported as ASCII files, by selecting the check box *Write Moments 0 to 6 in a file* and indicating a file suffix. The output files will be written into the directory selected in the following dialogue and called `Moment*.suffix`.

The following commands are available in the *Modify menu*:

- **Interpolate Dispersion**
  Linear interpolation of all selected spectra onto a common grid of dispersion values. This is necessary for most data operations such as computing line statistics, Fourier analysis, least-squares fitting and mode identification with the FPF method. In the dialogue window, three different options for the interpolation can be selected. In all three cases, the interpolated spectra will be written to a new data set.

  - Interpolate onto scale of first spectrum: all spectra will be interpolated onto the dispersion scale of the first spectrum of the currently selected spectra.
  - Choose a file to interpolate: interpolate onto the scale read from an ASCII file. In the dialogue window you can select in which column the dispersion values are listed.
  - Compute grid for interpolation: interpolate on a grid of equidistant dispersion values. The minimum, maximum and step values must be indicated by the user.

- **Convert Dispersion**
  Convert the dispersion scale of the selected spectra from Ångstrom to
  km s^{-1} or vice versa, dependent on the dispersion scale of the current
  spectra. The value of the zero-point for the conversion must be indicated
  in the dialogue window. The converted spectra will be written into a new
  data set.

- **Extract Line**
  Opens a dialogue for determining the position of a line profile using sigma
  clipping. This tool is especially useful when line moments have to be
  calculated for a time series where the wavelength position line profile
  shifts significantly due to pulsation. Since ideally, for the computation
  of the moments, the continuum should not be included, the line has to
  be extracted. This tool determines the position of the left and right line
  limits through a sigma clipping algorithm, which detects the continuum.

  Description of the extract line dialogue (see Figure 6):

  The *Times of Observations Box* shows the list of times of the selected
  spectra. Clicking on a time will show the according spectrum in the
  Current Spectrum Box.

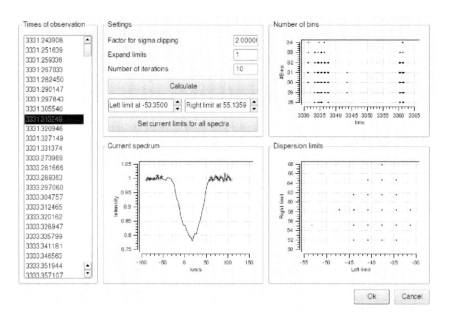

Figure 6: Screenshot of the dialogue for extracting a spectral line.

In the *Settings Box*, the sigma-clipping factor and the number of iterations can be indicated and must be adapted for different data sets. Sigma clipping iteratively removes outliers of a Gaussian distribution. In this case, the sigma clipping algorithm tries to find the pixels belonging to the continuum and thus to determine the limits of the line. These limits can be expanded with a number of pixels indicated with *Expand limits*.

Alternatively, the limits can be set at certain dispersion values indicated in the spin boxes. The limits can be changed for each spectrum individually or applied to all spectra when clicking on *Set current limits for all spectra*.

After clicking on *Calculate*, the position of the pixels detected as continuum are marked as red crosses in the *Current Spectrum Box*. The two plots on the right-hand side show information about all spectra of the time series in order to check the overall results. The top plot shows the number of bins detected as line of each spectrum. The bottom plot shows the left and right dispersion limits of each spectrum. Outliers in this plot (marked red in the plots and in the *Times of Observation Box*) can indicate that the line position has not been detected properly in some spectra. When clicking on *OK*, the extracted line will be written into a new data set.

- **Shift Dispersion**
  Shift the zero-point of the dispersion of the selected spectra with a fixed value (positive or negative). The shifted spectra will be written into a new data set.

- **Subtract Mean**
  Subtract the temporal mean from all selected spectra, i.e., compute the difference of each spectrum from the mean spectrum. The dispersion scale of all spectra must be interpolated on each other to use this function. The new spectra will be written into a new data set.

- **Add Noise**
  Add white Gaussian noise to the selected spectra. The continuum SNR must be indicated in the dialogue window. The new spectra will be written into a new data set.

4.1.3 Spectrum Box

This box shows a list of the currently selected spectrum in the *Time Series Box*. It consists of two columns, dispersion and intensity. The dispersion can be in units of Ångstrom or km s^{-1}, dependent on the selected data set. Multiple bins can be selected by clicking with the left mouse button on several items in the list while pressing the Ctrl-key or the Shift-key. All items can be selected by pressing *Select All*. Selected items are displayed in the plot window as red crosses.

The following commands are available in the menu Data:

- **Edit Data**
  Opens a table of dispersion values and intensity in a new window with the possibility to edit these values. Modifications can be written to the current spectrum.

- **Remove Selection**
  Remove the selected bins from the spectrum/data set. Use this function to remove bad pixels with deviating intensities from the spectra. After removal of a pixel, interpolation of the spectra onto a common velocity grid might be necessary.

4.1.4 Plot Window

The plot window shows the currently selected spectrum with selected wavelength bins, a time series of moments, or the statistics of a time series of spectra (mean, standard deviation).

For more information about the plot window, we refer to p. 28.

4.2   Fourier Analysis

With this module, a discrete Fourier transform (DFT) can be computed to search for periodicities in the data set selected in the *Data Sets Box* of the *Data Manager Tab*. The data can consist of a time series of spectra (two-dimensional) or of a time series of moments (one-dimensional). For the latter, we refer to the photometry manual (see p. 96). To compute a Fourier analysis for a time series of spectra, you must indicate the dispersion range (in Ångstrom or km s^{-1}) that should be taken into account, the frequency range, and what the calculations are based on (pixel-by-pixel line profile or moments). The Fourier spectrum is displayed in the plot window and saved as data set in the *List of calculations*. A screenshot of the *Fourier Tab* is displayed in Figure 7.

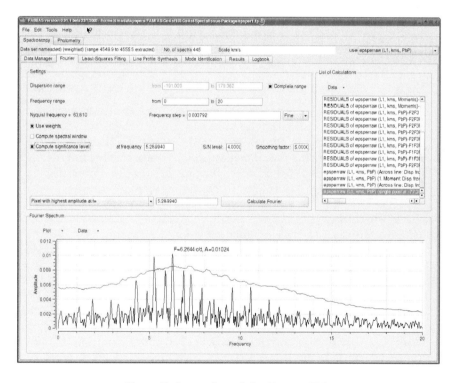

Figure 7: Screenshot of the *Fourier Tab*.

4.2.1   Settings Box

In this box, the settings for the Fourier analysis are defined.

- **Dispersion range**
  Minimum/Maximum values of the dispersion range in Ångstrom or km s^{-1}, dependent on the input data. The range specifies which wavelength bins of the spectrum will be taken into account for the computation of the Fourier spectrum. The Complete range is selected if the corresponding box is checked.

- **Frequency range**
  Minimum/maximum values of the frequency range. The Fourier spectrum will be computed from the minimum to the maximum value.

- **Nyquist frequency**
  Estimate of the Nyquist frequency (mean sampling frequency). For non-equidistant time series, a Nyquist frequency is not uniquely defined. In this case, the Nyquist frequency is approximated by the inverse mean of the time-difference of consecutive measurements by neglecting large gaps.

- **Frequency step**
  Step size (resolution) of the Fourier spectrum. Three presets are available: Fine $(\equiv (20\Delta T)^{-1})$, Medium $(\equiv (10\Delta T)^{-1})$, and Coarse $(\equiv (5\Delta T)^{-1})$. The corresponding step size depends on the temporal distribution of the measurements, i.e., the time difference ΔT of the last and first measurement. It is recommended to select the fine step size to ensure that no frequency is missed. The step value can be edited if desired.

- **Use weights**
  If the box is checked, the weight indicated for each spectrum is taken into account in the Fourier computations. Otherwise, all weights are assumed to have equal values.

- **Compute spectral window**
  If the box is checked, a spectral window of the current data set is computed. A spectral window shows the effects of the sampling of the data on the Fourier analysis and thus permits to estimate aliasing effects. The spectral window is computed from a Fourier analysis of the data taking the times of measurements and setting all measurement intensities to the value 1. The shape of the spectral window does not depend on the selected dispersion range and should be plotted for a frequency range that is symmetric around 0 for visual inspection.

- **Compute significance level**
  If the box is checked, the significance level at a certain frequency value is computed and shown in the plot window as a red line. The following parameters can be set:

  - **Frequency**
    Frequency value of the peak of interest. The data will be pre-whitened with this frequency and the significance level will be computed from the pre-whitened Fourier spectrum.

  - **S/N level**
    Multiplicity factor of the signal-to-noise level. The displayed significance level will be multiplied by this factor.

  - **Box size**
    Box size b for the computation of the noise-level in units of the frequency. The displayed significance level is computed from the running mean of the pre-whitened Fourier spectrum. For each frequency value F, the noise level is calculated from the mean of the range $[F - b/2, F + b/2]$.

  You must choose what data the calculations are based on and then press *Calculate Fourier*. The significance level will be shown as a red line in the plot window together with the Fourier spectrum of the data (blue line).

  This option cannot be selected when computing a Fourier spectrum across the line profile. In this case, no signal-to-noise criterion (e.g., significance of a peak when SNR ≥ 4) can be applied, because the computed Fourier spectrum is an average of all Fourier spectra across the line profile. To determine the significance of a frequency peak across the line profile, one should use the function *Pixel with highest amplitude at f=* in combination with *Compute significance level*. By doing so, only the dispersion bin having the largest amplitude of the indicated frequency is taken into account for the computation of its SNR.

- **Calculations based on**
  Defines what the calculation of the Fourier analysis is based on. The following settings are possible:

  - **Pixel-by-pixel (1D, mean Fourier spectrum)**
    Computes a Fourier spectrum which is the mean of all Fourier spectra across the selected dispersion range. The resulting Fourier spectrum is therefore one-dimensional with frequency on the x-axis and mean amplitude on the y-axis. The signal-to-noise ratio of a peak

cannot be determined since a frequency can have different amplitudes across the line profile. For this, use the option *Pixel with
highest amplitude at f=* in combination with *Compute significance
level.*

- **Pixel-by-pixel (2D, only export)**
  Computes a Fourier spectrum for each pixel (= bin) across the selected dispersion range. The output is a two-dimensional Fourier
  spectrum where the amplitude is a function of frequency and dispersion. Due to the generally large data size of such a Fourier
  spectrum (some megabytes), it can only be exported to an ASCII
  file. A contour plot of these data can easily be created by the user,
  e.g., with the program gnuplot with the commands set pm3d map
  and splot.

- **Pixel with highest amplitude at f =**
  Computes a Fourier spectrum at the pixel where the given frequency
  has the highest amplitude. The purpose of this task is to determine
  the significance of a frequency peak in a line profile. You must
  indicate a frequency value to carry out this operation. This task
  computes for each pixel across the selected dispersion range a Fourier
  spectrum and determines at which position in the profile the given
  frequency has the highest amplitude.

- **Equivalent width**
  Computes the equivalent width of the line profile (inside the indicated dispersion range) and calculates its Fourier spectrum.

- **1st moment (radial velocity)**
  Computes the first moment $<v^1>$ of the line profile (inside the
  indicated dispersion range) and calculates its Fourier spectrum.

- **2nd moment (variance)**
  Computes the second moment $<v^2>$ of the line profile (inside the
  indicated dispersion range) and calculates its Fourier spectrum.

- **3rd moment (skewness)**
  Computes the third moment $<v^3>$ of the line profile (inside the
  indicated dispersion range) and calculates its Fourier spectrum.

- **4th moment**
  Computes the fourth moment $<v^4>$ of the line profile (inside the
  indicated dispersion range) and calculates its Fourier spectrum.

- **5th moment**
  Computes the fifth moment $<v^5>$ of the line profile (inside the
  indicated dispersion range) and calculates its Fourier spectrum.

- **6th moment**
  Computes the sixth moment $<v^6>$ of the line profile (inside the indicated dispersion range) and calculates its Fourier spectrum.

- **Calculate Fourier**
  Computes the discrete Fourier transform (DFT) according to your settings and displays it in the plot window as a blue line. The mean of the time series is automatically shifted to zero before the Fourier transform is computed. The peak having highest amplitude in the given range is marked in the plot window. A dialogue window reports the frequency having the highest amplitude in the selected frequency range and asks if it should be added to the frequency list of the *Least-Squares Fitting Tab*.

4.2.2 List of Calculations

Previous Fourier calculations can be selected from the list. Each computed Fourier spectrum is saved and listed here. If a project is saved, the list of computed Fourier spectra is also saved but compressed to decrease the project file size (only extrema are saved). The following operations are possible via the *Data Menu*:

- **Remove Data Set**
  Removes the currently selected data set from the list.

- **Rename Data Set**
  Renames the currently selected data set.

- **Export Data Set**
  Exports the currently selected data set to an ASCII file having the following three-column format: frequency, amplitude, power.

4.2.3 Fourier Spectrum Plot

Shows the most recently computed Fourier analysis or the selection from the list of calculations. The Fourier spectrum is shown as a blue line, the significance level, if included, is shown as a red line. The frequency and amplitude of the peak having the highest frequency are indicated.

For more information about the plot window, we refer to p. 28.

4.3   Least-Squares Fitting

This module provides tools to compute a non-linear multi-periodic least-squares fit of a sum of sinusoidals to your data. The fitting can be applied for every bin of the spectrum separately (pixel-by-pixel) or for the different line moments. The fitting formula is

$$Z + \sum_i A_i \sin\left[2\pi(F_i t + \phi_i)\right]. \tag{6}$$

Here, Z is the zero-point, and A_i, F_i, and ϕ_i are respectively amplitude, frequency and phase (in units of 2π) of the i-th frequency.

The least-squares fit is carried out with the Levenberg-Marquardt algorithm. For a given set of frequencies, either their zero-point, amplitude and phase can be optimised (*Calculate Amplitude & Phase*), or additionally also the frequency value itself (*Calculate All*). The latter is only available for one-dimensional time series (i.e., the line moments). The data can be pre-whitened with the computed fit and written to the *Data Sets Box* of the *Data Manager Tab*. A screenshot of the *Least-Square Fitting Tab* is displayed in Figure 8.

Before a mode identification can be carried out, a least-squares fit to the data must be calculated. In order to apply the Fourier parameter fit method, the fit must be based on the pixel-by-pixel values. To apply the moment method, the fit must be based on the first moment.

4.3.1   Settings

Defines the settings for the calculation of the least-squares fit.

- **Dispersion range**
  Minimum/Maximum values of the dispersion range in Ångstrom or km s^{-1} (dependent on the input data). The range specifies which wavelength bins of the spectrum will be taken into account for the computation of the least-squares fit.

- **Use weights**
  If this box is checked, the weight indicated for each spectrum is taken into account in the least-squares fit. Otherwise, all weights are assumed to have equal values.

- **Pre-whiten data**
  If this box is checked, the data will be pre-whitened with the computed least-squares fit and written into a new data set. If the calculations are based on pixel-by-pixel, a new time series of spectra will be created. In

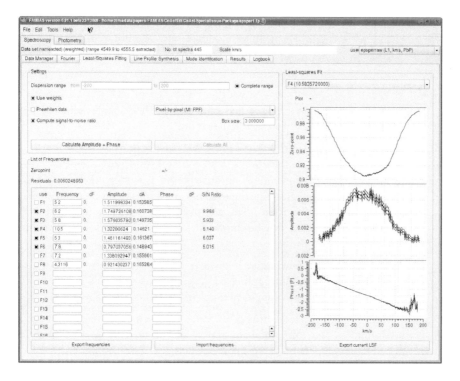

Figure 8: Screenshot of the *Least-Squares Fitting Tab*.

this case, the zero-point profile will not be taken into account for the pre-whitening to preserve the mean shape of the line profile. When a least-squares fit of a line moment is computed, the pre-whitened time series of moments is written into a new data set (one-dimensional time-series).

- **Calculations based on**

  This drop-down box defines what the computation of the least-squares fit is based on. The following settings are possible:

  - **Pixel-by-pixel (MI: FPF)**

    For each pixel (= dispersion bin) across the selected dispersion range, a separate least-squares fit is computed by improving zero-point, amplitude, and phase. For this option, the frequency value cannot be improved. The results of the fit for each frequency are displayed in the plot window. The integral of the amplitude across the line in the indicated dispersion range is written to the frequency list.

This option has to be chosen if the Fourier parameter fit mode identification method should be applied. The computed least-squares fits can be imported from the *Mode Identification Tab*.

– **Equivalent width**
Computes the equivalent width of the line profile (inside the indicated dispersion range) and calculates a least-squares fit. The results are written to the frequency list.

– **1st moment (radial velocity, MI: moment)**
Computes the first moment $<v^1>$ of the line profile (inside the indicated dispersion range) and calculates a least-squares fit. This option has to be chosen if the moment method should be applied for the mode identification. The computed least-squares fit and time series of moments can be imported from the *Mode Identification Tab*.

– **2nd moment (variance)**
Computes the second moment $<v^2>$ of the line profile (inside the indicated dispersion range) and calculates a least-squares fit.

– **3rd moment (skewness)**
Computes the third moment $<v^3>$ of the line profile (inside the indicated dispersion range) and calculates a least-squares fit.

– **4th moment**
Computes the fourth moment $<v^4>$ of the line profile (inside the indicated dispersion range) and calculates a least-squares fit.

– **5th moment**
Computes the fifth moment $<v^5>$ of the line profile (inside the indicated dispersion range) and calculates a least-squares fit.

– **6th moment**
Computes the sixth moment $<v^6>$ of the line profile (inside the indicated dispersion range) and calculates a least-squares fit.

• **Compute signal-to-noise ratio**
Computes the amplitude SNR of each selected frequency and displays it in the list of frequencies. The noise is computed from the Fourier spectrum of the pre-whitened data. The *Box size* is the width of the frequency range which is taken into account for the calculation of the noise. For a box width of b, the noise of a given frequency F is the mean value of the Fourier spectrum of the residuals in the range $[F - b/2, F + b/2]$.

How the SNR is computed depends on the selected calculation basis. In the case of pixel-by-pixel, for each frequency, the dispersion bin where this frequency has the highest amplitude is determined. The SNR is derived from this bin alone. For the moments, the SNR is computed from the ratio of A_F and the noise of pre-whitened Fourier spectrum at the position of F.

- **Calculate Amplitude + Phase**
  Computes a least-squares fit with the Levenberg-Marquardt algorithm using the above mentioned fitting formula. The zero-point, amplitude and phase are calculated, whereas the frequency is kept fixed.

  If the computations are based on pixel-by-pixel, the determined (improved) values of zero-point, amplitude, and phase are plotted for each frequency in the plot window. The uncertainties are derived from the error matrix of the least-squares fitting algorithm. The residuals (=mean standard deviation of the residuals) and the integral of the amplitude across the selected dispersion range are written to the frequency list.

  For the moments, the following optimised values are written into the frequency list: the zero-point and its uncertainty, the standard deviation of the residuals, for each selected frequency its amplitude and phase, and their formal uncertainties derived from the error matrix of the least-squares fitting algorithm.

- **Calculate All**
  Computes a least-squares fit with the Levenberg-Marquardt algorithm using the above mentioned fitting formula. The zero-point, amplitude, phase and frequency are improved. This option cannot be selected for computing a least-squares fit across the profile (pixel-by-pixel). For the moments, the following optimised values are written to the frequency list: the zero-point and its uncertainty, the standard deviation of the residuals, the frequency, amplitude and phase and their formal uncertainties derived from the error matrix of the least-squares fitting algorithm.

4.3.2 List of Frequencies

The *List of Frequencies Box* displays the results of the least-squares fit. Frequencies that should be included in a least-squares fit can be entered in the column *Frequency* and selected by clicking on the check box in the column *Use*. The following values are shown in this box after a least-squares fit has been calculated:

- **Least-squares fit across the profile with the option Pixel-by-pixel**
  The mean standard deviation of the residuals (pre-whitened spectra) across the selected dispersion range (Residuals) and, for each selected frequency, the integral of the amplitude across the dispersion range and its uncertainty are shown. If *Compute signal-to-noise ratio* has been selected, the amplitude SNR of each frequency is shown in the column SNR. This value refers to the SNR of the dispersion bin where the particular frequency has its highest amplitude.

- **Least-squares fit with the option Moments (Equivalent width and 1st through 6th moment)**
  The zero-point, its formal uncertainty and the standard deviation of the residuals are shown at the top. The improved values of frequency, amplitude and phase and their formal statistical uncertainties are shown in the list. The phase and its uncertainty is in units of 2π. The last column lists the SNR for each frequency (only shown when the box *Calculate signal-to-noise ratio* has been checked). The SNR is computed from the Fourier spectrum after pre-whitening with all selected frequencies. For each frequency, the assumed noise-level is computed from the mean amplitude around the frequency value with the box size indicated in the field *Calculate signal-to-noise ratio*.

- **Export frequencies**
  Exports all frequency, amplitude and phase values of the List of frequencies to an ASCII file. The file format is compatible with the program Period04 (Lenz & Breger 2005[1]).

- **Import frequencies**
  Imports an ASCII list of frequencies having the following four-column format separated by tabulators: frequency counter, frequency value, amplitude, phase (see example below). Bracketed values are unchecked frequencies. This format is compatible with the program Period04 (Lenz & Breger 2005[1]).

```
F1    (5.2861     0.029179815     0.534 )
F2     6.2566     0.017759398     0.7461502
F3     5.885284   0.029203887     0.47617591
F4    (10.583572  0.022958049     0.55097456 )
```

---

[1] http://www.univie.ac.at/tops

4.3.3   Least-Squares Fit plot

The plot window displays zero-point, amplitude and phase and their uncertainties of the current least-squares fit across the line profile (only active when pixel-by-pixel was selected for the calculations). The frequency can be selected in the combo box at the top.

- **Export current LSF**
  Export the current least-squares fit across the line profile to ASCII-files. You must indicate a file name with an extension (like name.ext). For each frequency x, a separate output file, called name_Fx.zap is created. The files consist of the following columns: dispersion value, zero-point, standard deviation of the zero-point, amplitude, standard deviation of the amplitude, phase (in units of 2π), standard deviation of the phase.

For more information about the plot window, we refer to p. 28.

4.4   Line Profile Synthesis

This module can be used to compute a time series of synthetic line profiles of a multi-periodically radially or non-radially pulsating star. The synthetic line profiles are written as a new data set to the *Data Manager Tab*. The Fourier parameter fit method in FAMIAS uses the same implementation for the computation of the synthetic line profiles. A screenshot of the *Line Profile Synthesis Tab* is displayed in Figure 9.

4.4.1   Theoretical background

We now briefly describe the approach for computing the line profiles. For a more detailed description, we refer to Zima (2006). The following is slightly modified from this publication.

We assume that the displacement field of a pulsating star can be described by a superposition of spherical harmonics. Our description of the Lagrangian displacement field is valid in the limit of slow rotation taking the effects of the Coriolis force to the first order into account (Schrijvers et al. 1997). Since deviations from spherical symmetry due to centrifugal forces are ignored, our formalism is reliable only for pulsation modes whose ratio of the rotation to the angular oscillation frequencies $\Omega/\omega < 0.5$ (Aerts & Waelkens 1993). This limitation excludes realistic modeling of rapidly rotating stars and low-frequency g-modes. For higher frequency p-modes, such as observed in many δ Scuti and β Cephei stars, the given criterion is fulfilled and a correct treatment is provided.

The intrinsic line profile is assumed to be a Gaussian. This is a good approximation for strong spectral lines of metals where the rotational broadening dominates over other line-broadening mechanisms. A distorted profile is computed from a weighted summation of Doppler shifted profiles over the visible stellar surface. Additionally, we take into account a parametrised variable equivalent width due to temperature and brightness variations across the stellar surface.

We assume an unperturbed stellar model to be spherically symmetric, in hydrostatic equilibrium, and unaffected by a magnetic field or rotation. The position of a mass element of such a star can be written in spherical coordinates (r, θ, ϕ) defined by the radial distance to the stellar centre r, the co-latitude $\theta \in [0, \pi]$, i.e., the angular distance from the pole, and the azimuth angle $\phi \in [0, 2\pi]$. Any shift of a mass element from its equilibrium position is given by the Lagrangian displacement vector $\boldsymbol{\xi} = (\xi_r, \xi_\theta, \xi_\phi)$. This displacement modifies the initial pressure p_0, the density ϱ_0, and the gravitational potential Φ_0 as a function of r, θ, ϕ, and the time t. The linear, adiabatic perturbations of these parameters are governed by the four equations of hydrodynamics, i.e., Poisson's equation, the equation of motion, the equation of continuity, and the energy equation, which translates into the condition for adiabacity in the absence of non-adiabatic effects in the stellar envelope.

This set of differential equations is solved by assuming that all perturbed quantities depend on $Y_\ell^m(\theta, \phi)\, e^{i\omega t}$, where $Y_\ell^m(\theta, \phi)$ denotes the spherical harmonic of degree ℓ and of azimuthal order m, ω is the angular pulsation frequency, and t the time. The spherical harmonic can be written as

$$Y_\ell^m(\theta, \phi) \equiv N_\ell^m P_\ell^{|m|}(\cos\theta)\, e^{im\phi}. \tag{7}$$

Here, $P_\ell^{|m|}$ denotes the associated Legendre function of degree ℓ and azimuthal order m, given by

$$P_\ell^m(x) \equiv \frac{(-1)^m}{2^\ell \ell!}(1-x^2)^{\frac{m}{2}}\frac{d^{\ell+m}}{dx^{\ell+m}}(x^2-1)^\ell, \tag{8}$$

and

$$N_\ell^m = (-1)^{\frac{m+|m|}{2}}\sqrt{\frac{(2\ell+1)}{4\pi}\frac{(\ell-|m|)!}{(\ell+|m|)!}} \tag{9}$$

is a normalisation constant. The definition of N_ℓ^m changes from author to author, which must be taken into account when comparing derived amplitudes. *In our formalism a positive value of m denotes a pro-grade mode, i.e., a wave propagating in the direction of the stellar rotation around the star.*

We model uniform stellar rotation, including first-order corrections due to the Coriolis force, which gives rise to toroidal motion. The resulting displacement field in the case of a slowly rotating non-radially pulsating star cannot be described by a single spherical harmonic anymore. It consists of one spheroidal and two toroidal terms, which only have a horizontal component, and is given for an angular frequency ω in the stellar frame of reference and a time t by

$$\begin{aligned}\boldsymbol{\xi} = \sqrt{4\pi}\Bigg[&a_{s,\ell}\left(1, k\frac{\partial}{\partial\theta}, k\frac{1}{\sin\theta}\frac{\partial}{\partial\phi}\right)Y_\ell^m(\theta,\phi)\,e^{-i\omega t}\\ &+a_{t,\ell+1}\left(0, \frac{1}{\sin\theta}\frac{\partial}{\partial\phi}, -\frac{\partial}{\partial\theta},\right)Y_{\ell+1}^m(\theta,\phi)\,e^{-i(\omega t+\frac{\pi}{2})}\\ &+a_{t,\ell-1}\left(0, \frac{1}{\sin\theta}\frac{\partial}{\partial\phi}, -\frac{\partial}{\partial\theta},\right)Y_{\ell-1}^m(\theta,\phi)\,e^{-i(\omega t-\frac{\pi}{2})}\Bigg]\end{aligned} \tag{10}$$

(Martens & Smeyers 1982, Aerts & Waelkens 1993, Schrijvers et al. 1997). Note that the term proportional to $Y_{\ell-1}^m$ is not defined for radial and sectoral modes. Here, $a_{s,\ell}$ denotes the amplitude of the spheroidal component of the displacement field, whereas $a_{t,\ell+1}$ and $a_{t,\ell-1}$ are the corresponding amplitudes of the toroidal components. We neglect the first order correction of the amplitude a_s due to rotation, whereby the amplitudes of the toroidal terms can be approximated by the following relations:

$$a_{t,\ell+1} = a_{s,\ell} \frac{\Omega}{\omega} \frac{\ell - |m| + 1}{\ell + 1} \frac{2}{2\ell + 1}(1 - \ell k),$$

$$a_{t,\ell-1} = a_{s,\ell} \frac{\Omega}{\omega} \frac{\ell + |m|}{\ell} \frac{2}{2\ell + 1}\left(1 + (\ell + 1)k\right). \tag{11}$$

The factor $\sqrt{4\pi}$ in Eq. (10) is introduced in order to scale the normalisation $\sqrt{4\pi}N_0^0 = 1$, such that a_s represents the fractional radius variation for radial pulsation.

The ratio of the horizontal to vertical amplitude, which attains quite different values for p- and g-modes, can be approximated by the following relation in the limit of no rotation

$$k \equiv \frac{a_h}{a_s} = \frac{GM}{\omega^2 R^3} \tag{12}$$

where a_h and a_s are the horizontal and vertical amplitude, G is the gravitational constant, M is the stellar mass, and R is the stellar radius.

We assume that the intrinsic line profile is a Gaussian, which may undergo equivalent width changes due to temperature variations. The distorted line profile is calculated from an integration of an intrinsic profile over the whole visible stellar surface, which - for computational purposes - numerically results in a weighted summation over the surface grid.

We define the intrinsic Gaussian profile in a surface point having the line-of-sight velocity V as

$$I(v, T_{\text{eff}}, \log g) = \left(1 + \frac{\delta F}{F}\right)\left[1 - \frac{W_{\text{int}}(T_{\text{eff}})}{\sigma\sqrt{\pi}}e^{-(\frac{V-v}{\sigma})^2}\right]. \tag{13}$$

Here, v is the velocity across the line profile, $\delta F/F$ takes the surface flux of the emitting segment into account, $W_{\text{int}}(T_{\text{eff}})$ is the equivalent width as a function of the effective temperature (see Eq. (14)); and σ is the width of the intrinsic profile. The distorted line profile is calculated by summation over all visible segments on the surface grid of (θ, ϕ) weighted over the projected surface.

The response of a line's equivalent width to local temperature changes is dependent on the involved element, its excitation, and the temperature in the zone where the line originates. In order to take this effect into account, a variable equivalent width of the intrinsic line profiles must be considered for calculating the distorted profile. Since there is no phase shift between $\delta W_{\text{int}}(T)$ and δT, we can write, following Schrijvers & Telting (1999),

$$W_{\text{int}}(T_{\text{eff}}) = W_0(1 + \alpha_W \delta T_{\text{eff}}), \tag{14}$$

where α_W is a parameter denoting the equivalent width's linear dependence on δT_{eff}, which can be approximated for $\delta T_{\text{eff}} \ll 1$. In FAMIAS, this parameter is denotes as $d(EW)/d(T_{\text{eff}})$.

For calculating the local temperature, surface gravity, and flux variations, we closely followed Balona (2000) and Daszyńska-Daszkiewicz et al. (2002). Since the flux variation $\delta F/F$ is mainly a function of T_{eff} and $\log g$, we can write in the limit of linear pulsation theory

$$\frac{\delta F}{F} = \alpha_T \frac{\delta T_{\text{eff}}}{T_{\text{eff}}} + \alpha_g \frac{\delta g}{g} =$$
$$= \frac{\delta R}{R_0}\left[\alpha_T f \frac{1}{4} e^{i\psi_f} - \alpha_g\left(2 + \frac{3\omega^2}{4\pi G <\rho>}\right)\right], \tag{15}$$

where α_T and α_g given by

$$\alpha_T = \left(\frac{\partial \log F}{\partial \log T_{\text{eff}}}\right)_g \text{ and } \alpha_g = \left(\frac{\partial \log F}{\partial \log g}\right)_{T_{\text{eff}}} \tag{16}$$

are partial derivatives of the flux, which can be calculated from static model atmospheres for different passbands. Here, R_0 is the unperturbed radius, G denotes the gravitational constant, $<\rho>$ is the mean density of the star, f the absolute value of the complex $f_R + i f_I$, and ψ_f the phase lag of the displacement between the radius and temperature eigenfunctions. Then f describes the ratio of flux to radius variations, which can be transformed into the ratio of temperature to radius variations due to the fact that the flux is proportional to T^4.

4.4.2 Stellar Parameters

In this box, the global stellar parameters that should be used for the computation of the synthetic line profiles are defined.

- **Radius**
  Stellar radius in solar units. In combination with the stellar mass, this parameter determines the k-value of the pulsation mode, i.e., the ratio of the horizontal to vertical displacement amplitude.

- **Mass**
  Stellar mass in solar units. In combination with the stellar radius, this parameter determines the k-value of the pulsation mode, i.e., the ratio of the horizontal to vertical displacement amplitude.

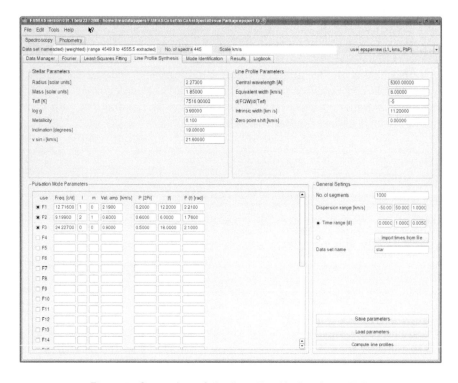

Figure 9: Screenshot of the *Line Profile Synthesis Tab.*

In FAMIAS, the following non-linear limb darkening law, described by Claret et al. (2000), is used to determine the brightness of the surface elements as a function of the line-of-sight angle α:

$$\frac{I(\mu)}{I(1)} = 1 - \sum_{k=1}^{4} a_k (1 - \mu^{\frac{k}{2}}). \tag{17}$$

Here, $I(\mu)$ is the specific intensity on the stellar disk at a certain line-of-sight angle θ with $\mu = \cos\theta$ and a_k is the k-th limb darkening coefficient.

The limb darkening coefficients are determined through the values of T_{eff}, $\log g$, and metallicity by bi-linear interpolation in a precomputed grid (Claret et al. 2000).

- **Teff**
  Effective temperature of the stellar surface in Kelvin.

- **log g**
  Value of the logarithm of the gravity at the stellar surface in c.g.s. units.

- **Metallicity**
  Stellar metallicity [m/H] relative to the sun.

- **Inclination**
  Angle between the line of sight and the stellar rotation axis, which is assumed to be the symmetry axis for pulsation, in degrees.

- **v sin i**
  Projected equatorial rotational velocity in km s^{-1}. The model assumes rigid rotation.

4.4.3 Line Profile Parameters

In this box, parameters of the synthetic line profile are defined.

- **Central wavelength**
  Central wavelength of the line profile in units of Ångstrom. This parameter determines the limb darkening coefficients, which are linearly interpolated in precomputed grids using the formalism by Claret (2000).

- **Equivalent width**
  Equivalent width of the line profile in km s^{-1}.

- **d(EW)/d(Teff)**
  Ratio between the equivalent width variations of the local intrinsic Gaussian line profile and the local temperature variations. This parameter can have positive as well as negative values (in the latter case the equivalent width decreases with increasing temperature). In combination with the non-adiabatic parameter f, this parameter controls the temporal equivalent width variations of the line profile.

- **Intrinsic width**
  Width of the intrinsic Gaussian line profile in km s^{-1}. This is the width of the line profile, unbroadened by stellar rotation and pulsation.

- **Zero-point shift**
  Shift of the line profile with respect to zero Doppler velocity in km s^{-1}. The synthetic line profiles are computed for the assumption that the barycentre of the line profile is at zero Doppler velocity.

4.4.4   Pulsation Mode Parameters

In this list, the parameters of the pulsation modes are defined.

- **Use**
  If a box is checked, the corresponding pulsation mode is taken into account for the computation of the synthetic line profiles.

- **Frequency**
  Value of the pulsation frequency in d^{-1} in the observer's frame of reference.

- **Degree ℓ**
  Spherical degree ℓ of the pulsation mode ($\ell \geq 0$).

- **Order m**
  Azimuthal order m of the pulsation mode ($|m| \leq \ell$). A positive value of m denotes a pro-grade pulsation mode.

- **Vel. Amp.**
  Velocity amplitude of the pulsation mode in km s^{-1}. The amplitude is normalised in such a way that it represents the intrinsic velocity for a radial pulsation mode.

- **P**
  Phase ϕ of the pulsation mode in units of 2π.

- **$\|f\|$**
  Absolute value of the complex non-adiabatic parameter f. For a definition, we refer to Eq. (30) on p.103. In combination with the parameter $d(EW)/d(T\!e\!f\!f)$ this parameter controls the equivalent width variations of the line profile.

- **P (f)**
  Phase lag ψ_f between the radius and temperature eigenfunctions, in units of radians.

4.4.5   General Settings

In this box, some general parameters for the computation of the synthetic line profiles can be defined.

- **No. of segments**
  Total number of segments (visible and invisible) on the stellar surface.

The segments are uniformly distributed across the surface, i.e., each segment covers approximately the same surface area. The segments lie on a spiral that has its endpoints at the poles of the sphere. At each segment, a local intrinsic Gaussian profile is defined and shifted by the local Doppler velocity. The overall synthetic line profile is computed by summing up over all visible local profiles. The higher the number of segments is, the better the precision of the computation gets at the cost of computational speed (linear increase).

- **Dispersion range**
  These values define the dispersion grid in Doppler velocity (km s^{-1}) for the computation of the synthetic line profiles. A minimum, maximum, and a step value must be indicated. Internally, a fixed step size of 1 km s^{-1} is taken, and the minimum and maximum limits expanded by 20 km s^{-1} to ensure that the profile is computed correctly at the limits. These synthetic profiles are then linearly interpolated onto the grid defined by the dispersion range.

- **Time range**
  Defines the minimum, maximum, and step values of the grid of times that should be used for the computation of the line profiles.

- **Import times from file**
  This allows the user to import a file that contains a list of time values that should be used for the line profile computation.

- **Data set name**
  Name of the data set of synthetic line profiles that is written to the *Data Manager Tab*.

- **Save parameters**
  Saves the parameters you entered in this tab to a file.

- **Load parameters**
  Loads the parameters for computing synthetic line profiles from a file.

- **Compute line profiles**
  Computes the synthetic line profiles and writes them into a new data set to the *Data Manager Tab*.

4.5   Mode Identification

This module is dedicated to the spectroscopic mode identification with the Fourier parameter fit (FPF) method and the moment method. Its main functions are: importing the observational data for the mode identification (least-squares fit across the line profile or line moments), setting stellar and pulsational parameters, defining the free parameters for the optimisation (=mode identification), and setting the parameters of the optimisation procedure. The results of the mode identification are written to the *Results Tab* and to log-files on the disk. A screenshot of the *Mode Identification Tab* is displayed in Figure 10.

Observational data can be imported and the parameters to be optimised can be chosen. Two different approaches for the identification of pulsation modes are available: the FPF method (Zima 2006) and the moment method (Balona 1986a,b, 1987; Briquet & Aerts 2003). Both methods assume the following: oscillations in the limit of linearity (sinusoidal variations), slow rotation (neglecting second order rotational effects), a limb darkening law according to Claret (2000), a symmetric intrinsic line profile, which is a Gaussian for the FPF method, and a displacement field that can be described by a sum of spherical harmonics. The FPF method furthermore permits to model a variable equivalent line width caused by local temperature variations on the stellar surface.

Both methods rely on the fact that the bin-intensities across an absorption line profile vary with the period of the associated non-radial pulsation mode. Whereas the FPF method makes use of the intensity information of each dispersion bin across the line profile, the moment method uses integrated values across the profile. This is the main difference between the two methods and leads to a difference in the capability of identifying pulsation modes. For the FPF method there is in principle no upper limit for the identification of (ℓ, m), but a very small value of $v \sin i$ as well as a large pulsation velocity relative to the projected rotational velocity can make mode identification impossible. In the latter two cases, the moment method is better suited, but this method is only sensitive for low-degree pulsation modes with $\ell \leq 4$. In the way as they are implemented in FAMIAS, both methods take into account the uncertainties of the observations and the goodness of the fit (=mode identification) is expressed as a chi-square value. The optimisation is carried out using genetic optimisation. Such an approach permits to search for local minima, and consequently the global minimum, in a complex large multi-parameter space.

For excellent reviews of spectroscopic mode identification techniques we refer to Aerts & Eyer (2000), Balona (2000), and Mantegazza (2000).

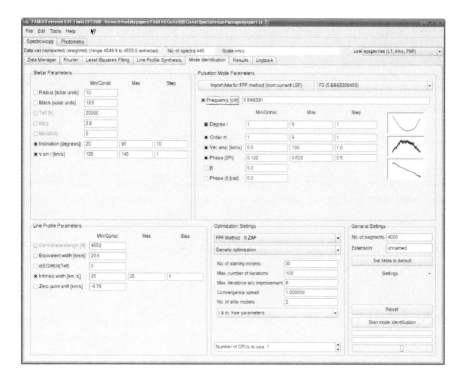

Figure 10: Screenshot of the *Mode Identification Tab*.

4.5.1 The FPF method

This method relies on the rotational broadening of a line profile and thus delivers good and reliable results for $v \sin i > 20$ km s^{-1}. The main assumptions of the models have been described above. For a more detailed description of this method, we refer to Zima (2006).

For each detected pulsation frequency and each dispersion bin across the line profile, a multi-periodic non-linear least-squares fit of sinusoids is computed (use the *Least-Squares Fitting Tab* of FAMIAS). This delivers the observational values of zero-point (Z_o), amplitude (A_o) and phase (P_o) as a function of the position in the line profile. These observational values are fitted with theoretical values derived from synthetic line profiles.

The FPF method comes in different flavours in FAMIAS, the main differences concerning the temporal distribution of the synthetic line profiles and the number of pulsation modes taken into account simultaneously. The FPF method makes use of the fact that the zero-point, amplitude and phase (ZAP) across the line profile depend on the (ℓ, m)-values of the associated pulsation modes. By comparing the theoretical values of ZAP with the observed ZAP-values, one can, in principle, determine the degree and azimuthal order of a pulsation mode.

The reduced χ_ν^2, which is regarded as goodness of the fit, is calculated from complex amplitudes in order to combine amplitude and phase information as follows

$$\chi_\nu^2 = \frac{1}{2n_\lambda - N} \sum_{i=1}^{n_\lambda} \left[\frac{(A_{R,i}^o - A_{R,i}^t)^2}{\sigma_{R,i}^2} + \frac{(A_{I,i}^o - A_{I,i}^t)^2}{\sigma_{I,i}^2} \right]. \tag{18}$$

Here, n_λ is the number of pixels across the profile, N is the number of free parameters, A^o and A^t denote observationally and theoretically determined values, respectively, $A_R = A_\lambda \cos\phi_\lambda$ and $A_I = A_\lambda \sin\phi_\lambda$ are the real and imaginary part of the complex amplitude, and σ is the observational error.

Since the amplitude and phase of a given wavelength bin are treated as independent variables, the variances are calculated from

$$\sigma_{R,\lambda}^2 = \sigma(A_\lambda)^2 \cos^2\phi_\lambda + \sigma(\phi_\lambda)^2 A_\lambda^2 \sin^2\phi_\lambda, \tag{19}$$

$$\sigma_{I,\lambda}^2 = \sigma(A_\lambda)^2 \sin^2\phi_\lambda + \sigma(\phi_\lambda)^2 A_\lambda^2 \cos^2\phi_\lambda. \tag{20}$$

4.5.2   Optimisation settings for the FPF method

In the drop-down menu *Select MI method*, the following selections are possible as optimisation settings:

- **Compute vsini, EW, intrinsic width, and velocity offset (fit Z)** With this setting, the pulsationally independent parameters $v\sin i$, the equivalent line width, the intrinsic width σ, and the Doppler velocity offset are determined from a fit of a rotationally broadened synthetic line profile to the observational zero-point profile. This method only provides reliable results if the line profile is not significantly broadened by pulsation. The determined values can be used as starting values for the mode identification.

- **FPF Method: fit AP**
  For each selected pulsation frequency, a single-mode displacement field and the corresponding line profiles are computed for 10 phase bins evenly distributed over one pulsation cycle. The theoretical values for AP are computed from a mono-periodic least-squares fit to these synthetic line profiles. A chi-square value is computed by taking into account the observed and theoretical Fourier parameters and their observational uncertainties (for details see Zima, 2006). The zero-point across the line profile,

which gives a strong constraint on $v \sin i$, the intrinsic line width and the equivalent width, is ignored in this case. Therefore, this option should only be chosen if already good constraints on these global parameters are known.

This method assumes that the different pulsation modes do not have a significant influence on each other's ZAP values. Such an assumption is valid if the ratio of the radial velocity amplitude to the projected rotational velocity for all frequencies is < 0.2. For higher values, the ZAP-values across the line profile might be distorted and impossible to model with a single-mode displacement field. In this case, the approach *FPF Method (complete time series): fit AP* (see below) or the moment method are better suited.

- **FPF Method: fit ZAP**
  This option is identical to *FPF Method: fit AP* with the exception that also the observed and theoretical zero-points are taken into account for computing the fits.

- **FPF Method (complete time series): fit AP**
  With this option, multi-periodicity and the complete series of observational times are considered for applying the FPF method. Synthetic line profiles are computed from a multi-mode displacement field, taking all selected pulsation modes into account. For each time step of the observed time series, one profile is computed. The theoretical AP across the profile are derived from a multi-periodic least-squares fit. Since multi-periodicity is considered for this method, it can in principle be applied to stars for which the radial velocity amplitude to the projected rotational velocity is of the order of 1.

  This method is computationally much slower since not only 10 synthetic profiles but the complete time series have to be modelled. The zero-point across the line profile, which gives a strong constraint on $v \sin i$, on the intrinsic line width and on the equivalent width, is ignored in this case. Therefore, this option should only be chosen if already good constraints on these global parameters are available.

- **FPF Method (complete time series): fit ZAP**
  This option is identical to *FPF Method (complete time series): fit AP* with the exception that also the observed and theoretical zero-points are taken into account for computing the fits.

4.5.3   Practical information for applying the FPF method

- The dispersion range of the least-squares fit (ZAP across the line profile) that is imported to the *Mode Identification Tab* should cover the range where the amplitude reaches significant values. In general, the continuum should therefore be excluded and the line wings can in many cases be excluded. The least-squares fit should include all significant frequencies - also combination and harmonic frequencies since they can have a significant effect on the Fourier parameters of the other pulsation frequencies. During the mode identification, combination and harmonic frequencies should not be set as free parameters unless one has a reason to assume that they are pulsation modes intrinsic to the star.

- The stellar parameters radius, mass, $T_{\rm eff}$, $\log g$, and metallicity should be quite well-known. Radius and mass can be set as variable during the fit and have an influence on the k-value (ratio of horizontal to vertical displacement amplitude) of the pulsation modes. The three other parameters determine the limb darkening coefficients and slightly affect the fitted $v \sin i$ and intrinsic line width.

- Before starting the mode identification, one should determine starting values for $v \sin i$, the intrinsic width, the equivalent width, and the velocity zero-point offset. This can be done by selecting *Compute vsini, EW, intrinsic width, and velocity offset (fit Z)* in the field *Optimisation Settings*. For this optimisation no pulsation mode should be selected and $v \sin i$, the equivalent width, the intrinsic width and the zero-point shift should be set as a variable in a reasonable range. This mode of optimisation fits a theoretical rotationally broadened line profile to the observational zero-point profile. Any pulsational broadening of the latter is neglected during the fit and can lead to an overestimation of $v \sin i$ or the intrinsic width.

- The most reliable results for the mode identification will be obtained when following an iterative scheme. In general, one can fix the equivalent width and the zero-point shift, once they have been determined with sufficient precision in the previous step. The values of $v \sin i$ and the intrinsic width should be set as variable in a range that is determined by the fit in the previous step (taking the chi-square values into account). The inclination should be set as variable in the complete possible (realistic) range between about 5 and 90 degrees taking a step of about 10 degrees. It does, in general, not make sense to set the lower range to 0 since this would imply infinitive equatorial velocity if $v \sin i > 0$ km s^{-1}. For each pulsation mode, a separate mode identification should be acquired first (using *FPF Method: fit AP* or *FPF Method: fit ZAP*). The degree, the azimuthal order and amplitude should be set to reasonable ranges.

- If the pulsation frequency has a significant amplitude in the least-squares fit of the first moment, its phase value can be used for the mode identification. In this case, the phase can be set as variable in the range $\phi_{<v^1>}$, $\phi_{<v^1>} + 0.5$ with a step of 0.5. If the frequency is not detected in the first moment, the phase value has to be set as free in the range between 0 and 1 with a step-size of ≤ 0.01. After a first run of the mode identification with ℓ and m as free parameters (see field *Optimisation Settings*), one normally has a constraint on the phase value ϕ and it should be set to ϕ, $\phi + 0.5$ with a step of 0.5 (different pulsation modes have their best fit at phase values that differ by half a period - this is due to the fact that we limit the inclination angle to a range between 0 and 90 degrees and not between 0 and 180 degrees). The further mode identification should be carried out by setting the search method to ℓ *and* m: *grid search*.

- Equivalent width variations of the line profile due to local temperature variations at the stellar surface can be taken into account by considering the parameters $\|f\|$ and *Phase (f)* (see Eq. (15)) in combination with the parameter $d(EW)/d(Teff)$ which can be positive or negative (see Eq. (14)). The parameter space is significantly enlarged by setting these parameters as variable, so it is important to already have some constraints on ℓ and m before attempting to fit the equivalent width variations.

- Multi-mode identification gives only good results if already some constraint about ℓ and m of the pulsation modes has been obtained. Otherwise the genetic optimisation algorithm may end up in a local minimum due to the large parameter space.

- The number of segments on the stellar surface (see field *General Settings*) should have a value of at least 1000. The lower this number, the lower the precision of the computations. For slowly rotating stars having low-degree modes ($\ell < 4$), a value between 1000 and 3000 in general is sufficient. For more rapidly rotating stars ($v \sin i > 50$ km s^{-1}) and high-degree modes, this value should be between 3000 and 10000.

4.5.4   The moment method

This method uses the first (radial velocity) and second (line width) moments of a line profile as a discriminator for mode identification. The version of the moment method we adopted, has been described in detail by Briquet & Aerts (2003) and has been slightly modified in FAMIAS. The complete time series of observed moments is fitted with theoretical moments to determine ℓ and m. The main assumptions for the computation of the theoretical moments have been described above. We take into account the uncertainties of the observed moments that can be computed numerically if the signal-to-noise ratio of the spectra is known. The observational uncertainties are used to compute a chi-square value which provides a statistical criterion for the significance of the mode identification.

The formalism for the calculation of the statistical uncertainty of the moments has been described on p. 34. The reduced χ_ν^2 goodness-of-fit value is computed from

$$\chi_\nu^2 = \frac{1}{2N} \sum_{i=1}^{N} \left[\left(\frac{<v^1>_o - <v^1>_t}{\sigma_{<v^1>_o}} \right)^2 + \left(\frac{<v^2>_o - <v^2>_t}{\sigma_{<v^2>_o}} \right)^2 \right], \qquad (21)$$

where N is the number of measurements of the time series, and the indices o and t denote observed and theoretical values, respectively.

To speed up the computations of the theoretical moments, a grid of integrals is precomputed for all possible ℓ and m-combinations for $0 \leq \ell \leq 4$ and all inclinations between 0 and 90°. This computation is performed once the mode identification has been started and may take a few minutes. The precomputed integrals depend on the limb darkening coefficients and the number of segments on the stellar surface. They are thus only recomputed in a subsequent mode identification if the latter parameter or the parameters *Teff, log g, Metallicity,* or *Central wavelength* have been modified.

4.5.5 Practical information for applying the moment method

- The dispersion range that is imported to the *Mode Identification Tab* should cover the complete range of the line profile (from continuum to continuum). The best approach would be to extract the line profile (see p. 36) and select the complete dispersion range when computing the least-squares fit. The least-squares fit should include all significant frequencies as well as combination and harmonic frequencies. During the mode identification, the latter should not be set as free parameters unless one has a reason to assume that they are pulsation modes intrinsic to the star.

- The stellar parameters radius, mass, T_{eff}, $\log g$, and metallicity should be quite well known. Radius and mass can be set as variable during the fit and have an influence on the k-value (ratio of horizontal to vertical displacement amplitude) of the pulsation modes. The three other parameters determine the limb darkening coefficients and thus mainly affect the fitted $v \sin i$ and intrinsic line width.

- The inclination should be set as variable in the complete possible (realistic) range between about 5 and 90 degrees taking a step of about 5 degrees. It does in general not make sense to set the lower range to 0 since this would imply infinite equatorial velocity if $v \sin i > 0$ km s^{-1}.

- After importing the line moments, the phase of each pulsation mode is set as free parameter between $\phi_{<v^1>}$, and $\phi_{<v^1>} + 0.5$ with a step of 0.5 (different pulsation modes have their best fit at phase values that differ by half a period. This is due to the fact that we limit the inclination angle to a range between 0 and 90 degrees and not between 0 and 180 degrees).

- To obtain the most reliable results with the moment method, one should set all detected pulsation frequencies (except combinations/harmonics) as free parameters during the fit. This is due to the fact that the complete time series of observed moments is fitted with theoretical moments. The number of segments on the stellar surface (see *General Settings Box*) should have a value of at least 1000. The lower this number, the lower the precision of the computations becomes. For slowly rotating stars having low-degree modes $(\ell \leq 4)$, a value between 1000 and 3000 is in general sufficient. For more rapidly rotating stars $(v \sin i \geq 50$ km s$^{-1})$ this value should be between 3000 and 5000.

- The chi-square values of the fits are derived numerically from the computation of the line moments taking into account the SNR of the spectra. If the SNR is not known for the single spectra, one can provide a mean SNR (in the box *Line Profile Parameters*). This mean value can be obtained by computing the inverse of the standard deviation at the continuum close to the line profile. In this case, the chi-square value is based on the assumption that all spectra have the same SNR and may thus not be reliable. If the SNR of each spectrum is known and listed in the *Data Manager*, one should select the option *Individual signal-to-noise ratio* in the box *Line Profile Parameters*.

4.5.6 Setting of parameters

During the mode identification several values can be set as fixed or free parameters. These values are listed in the boxes *Stellar Parameters, Pulsation Mode Parameters*, and *Line Profile Parameters*. If the check box associated with the parameter is unchecked, the parameter is fixed at a constant value during the optimisation. In this case a value must be entered in the input box of the column *Min/Const*. A parameter can be set to be variable (free) during the optimisation if the check box has been checked. In this case, two additional input boxes appear and values for the search range (Min, Max, Step) have to be entered. Some parameters, such as T_{eff} or $\log g$ cannot be set as free parameters since they determine the limb darkening coefficient. The more parameters are set as variable simultaneously and the finer the step, the larger the parameter space becomes. This must be taken into account when setting the optimisation parameters to avoid ending up in a local minimum (see Section 4.5.10 for more details).

4.5.7 Stellar Parameters

This box defines the global stellar parameters that should be used for the optimisation.

- **Radius**
  Stellar radius in solar units. In combination with the stellar mass, this parameter determines the k-value of the pulsation mode, i.e., the ratio of the horizontal to vertical displacement amplitude.

- **Mass**
  Stellar mass in solar units. In combination with the stellar radius, this parameter determines the k-value of the pulsation mode, i.e., the ratio of the horizontal to vertical displacement amplitude.

- **Teff**
  Effective temperature of the star in Kelvin. Together with the parameters *log g* and *Metallicity*, this parameter defines the limb darkening coefficients by linear interpolation in a precomputed grid (Claret et al. 2000).

- **log g**
  Value of logarithm of the gravity at the stellar surface. Together with the parameters *Teff* and *Metallicity*, this parameter defines the limb darkening coefficients by linear interpolation in a precomputed grid (Claret et al. 2000).

- **Metallicity**
  Stellar metallicity [m/H] relative to the Sun in logarithmic units. Together with the parameters *log g* and *Teff*, this parameter defines the limb darkening coefficients by linear interpolation in a precomputed grid (Claret et al. 2000).

- **Inclination**
  Angle between the line of sight and the stellar rotation axis in degrees.

The models assume that the axis of rotation is aligned with the symmetry axis of the pulsational displacement field.

- **v sin i**
  Projected equatorial rotational velocity $v \sin i$ in km s^{-1}.

4.5.8 Pulsation Mode Parameters

This box defines the parameters of the pulsation modes that should be identified. The observed data can be imported by clicking on the button. The imported frequencies can be selected with the combo box. Each frequency that should be taken into account for the mode identification must be selected by clicking on the check box next to the frequency value.

- **Button Import Data**
  Import the observational data for the mode identification. You must first compute a least-squares fit across the line profile (for FPF method) or the first moment (for moment method) in the *Least-Squares Fitting Tab*.

  - **For the moment method**
    Select in the combo box *Calculations based on* of the *Least-Squares Fitting Tab* the option *1st moment* and compute the least-squares fit by clicking on *Calculate Amplitude + Phase* or *Calculate All*. The import button in the *Pulsation Mode Parameters Box* of the *Mode Identification Tab* will now display *Import data for moment method*. After clicking, the spectra on which the least-squares fit was based, and the selected frequencies and their phases are imported and displayed in the *Pulsation Mode Parameters Box*. The frequencies can be selected from the combo box next to the import button.

  - **For the FPF method**
    Select in the combo box *Calculations based on* of the *Least-Squares Fitting Tab* the option *Pixel-by-pixel* and compute the least-squares fit by clicking on *Calculate Amplitude + Phase*. The import button in the *Pulsation Mode Parameters box* of the *Mode Identification Tab* will now display *Import data for FPF method*. After clicking, the parameters zero-point, amplitude, and phase across the line profile and the selected frequencies are imported and displayed in the *Pulsation Mode Parameters Box*. The frequencies can be selected from the combo box next to the import button.

- **Frequency**
  Value of the pulsation frequency as it was imported from the *Least-Squares Fitting Tab*. This value cannot be modified (only by importing a new least-squares fit).

- **Degree ℓ**
  Spherical degree ℓ of the pulsation mode ($\ell \geq 0$). Defines the search parameter space for the associated pulsation mode. The step size must have a value of ≥ 1.

- **Order m**
  Azimuthal order m of the pulsation mode ($|m| \leq \ell$). Defines the search parameter space for the associated pulsation mode. The step size must have a value of ≥ 1.

- **Vel. Amp.**
  Velocity amplitude of the pulsation mode in km s^{-1}. The amplitude is normalised in such a way that it represents the intrinsic velocity for a radial pulsation mode.

- **Phase**
  Phase ϕ of the pulsation mode in units of 2π.

- **$\|f\|$**
  Absolute value of the complex non-adiabatic parameter f. For a definition, we refer to Eq. (30) on p.103. In combination with the parameter $d(EW)/d(T\!e\!f\!f)$ this parameter controls the equivalent width variations of the line profile.

- **P (f)**
  Phase lag ψ_f between the radius and temperature eigenfunctions, in units of radians.

4.5.9 Line Profile Parameters

In this box, the parameters of the line profile are defined.

- **Central wavelength**
  Central wavelength of the line profile in units of Ångstrom. This parameter determines the limb darkening coefficients, which are linearly interpolated in precomputed grids using the formalism by Claret et al. (2000). The limb darkening coefficients slightly influence the derived values of $v \sin i$ and the intrinsic width, but generally have negligible effect on the mode identification.

- **Equivalent width**
  Equivalent width of the line profile in km s^{-1}.

- **d(EW)/d(Teff)**
  Ratio between the equivalent width variations of the local intrinsic Gaussian line profile and the local temperature variations. This parameter can have positive as well as negative values (in the latter case the equivalent width decreases with increasing temperature). For a definition see Eq. (14).

- **Intrinsic width**
  Width of the intrinsic Gaussian line profile in km s^{-1}.

- **Velocity offset**
  Offset of the line profile with respect to zero Doppler velocity in km s^{-1}. The synthetic line profiles are computed for the assumption that the barycentre of the line profile is at zero Doppler velocity. In general, this is not the case for the observed line profiles.

The following parameters are only available for the moment method.

- **Centroid velocity**
  Centroid velocity of the line profile. In ideal cases, this is the mean radial velocity ($<v^1>$) of the star. It is in any case best to use the zero-point of the least-squares fit to the first moment (which is automatically done in FAMIAS), especially if the time series consists only of few measurements or the radial velocity amplitude is large.

- **Mean signal-to-noise ratio**
  This value is used for the computation of the statistical uncertainties of the line moments if the SNR of the individual spectra is not known. In this case, the determined χ_ν^2-values might not be reliable if some individual spectra deviate strongly from this value.

- **Individual signal-to-noise ratio**
  If the SNR is known for each spectrum, this option should be chosen to determine the statistical uncertainties of the line moments. The values of the SNR can be imported with the spectra (additional column in the list of times) or computed in FAMIAS in the *Data Manager* → *Calculate* → *Compute Signal-To-Noise Ratio* (see p. 32).

4.5.10 Optimisation Settings

In this box, the settings for the optimisation procedure are defined. The optimisation is carried out with a genetic algorithm (Michalewicz 1996). These settings are crucial for the mode identification and must be chosen very carefully. The most important aspect is to avoid ending up in a local minimum. Since the computations of theoretical line profiles and moments is generally very time consuming, one must find a compromise between the coverage of the parameter space and CPU time efficiency. Although FAMIAS provides default values for different optimisation problems, the best way to proceed is trial-and-error, i.e., to test different optimisation settings and to proceed iteratively.

- **Select MI method**
  Selection of the mode identification method. See above for a description of the different possibilities. In the case of the moment method, only the option *Moment method* is available.

- **No. of starting models**
  Generation size during the genetic optimisation. Larger values for a larger parameter space.

- **Max. number of iterations**
  Stop criterion for the genetic optimisation. This number defines after how many iterations (=generations) the optimisation will stop.

- **Max. iterations w/o improvement**
  Stop criterion for genetic optimisation. The optimisation stops if no improvement of the best found model has been achieved after n iterations.

- **Convergence speed**
  Defines how quickly the algorithm is forced to converge. Value must be between 0 and 1. Higher values cause quicker convergence at the cost of parameter space exploration and thus precision.

- **No. of elite models**
  This number defines how many of the best models will be copied unaltered to the following generation. This parameter ensures quicker convergence.

- **ℓ & m: free parameters/grid search**
  Defines if the ℓ and m values are free parameters in the given range, or if they are subsequently fixed (grid search) while the other parameters are being optimised.

- **Number of CPUs to use**
  Number of processors that are used in parallel during the optimisation.

4.5.11 General Settings

- **No. of segments**
  Total number (visible + invisible) of segments on the stellar surface to compute the line profiles. Higher numbers provide higher accuracy but slower computational speed (linear dependence).

- **Extension**
  Extension of output and log files. The output directory can be chosen in the *Settings menu* (see below). The default output directory is the directory of the project file. During the mode identification, a log-file, called logMI.extension, is written to the disk. It contains a list of all computed models, their χ^2_ν-values and parameter values. After a mode identification has stopped, the results are written to the file bestFitsLog.extension. The best 20 fits (ASCII files and plots containing theoretical and observed zero-point, amplitude, and phase or moments) are written to files in the directory defined in *Settings* (see below).

- **Set fields to default**
  Set the optimisation, line and stellar parameters to default values dependent on the selected mode identification method. The proposed values are just a guidance and have to be adapted for many optimisation problems.

- **Settings**
  Opens a menu that provides the following functions: silent mode, save/load parameters, set output path for the logfiles and clear all fields of the *Mode Identification Tab*. The silent mode toggles the updating of the progress bars. For some optimisations, the computational performace decreases significantly, if the progress bars are updated.

- **Reset**
  Resets the previous optimisation. Must be pressed if a new optimisation procedure should be started.

- **Start mode identification**
  Starts the mode identification. The optimisation process can be stopped at any time by clicking again on this button. The results of the mode identification will be written into the *Results Tab*. After an optimisation process has stopped, again clicking this button will continue the optimisation at the stage where it stopped. To begin a new process, click the button *Reset*.

- **Progress bars and counter**
  The progress bars show the total progress and the progress in the current iteration. Below the bars, a counter gives the total number of computed models.

4.6 Results

This window shows the results of the current and previously derived spectroscopic mode identifications. A list of the parameters of the best fitting models, the fits of the theoretical models to the observations, and diagrams where the free parameters are plotted against the corresponding χ_ν^2-values are displayed. The results of previously performed mode identification process are logged. Once a mode identification has been started, this window is updated regularly with the actual status of the optimisation. A screenshot of the *Results Tab* is displayed in Figure 11.

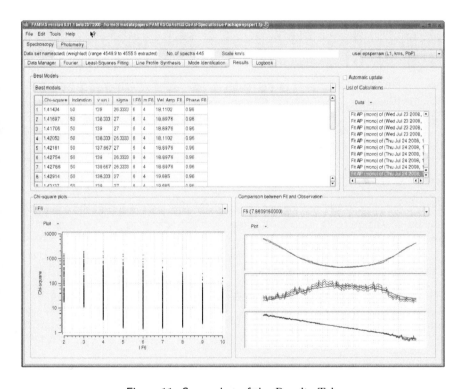

Figure 11: Screenshot of the *Results Tab*.

Press the button *Update* to update the list of best models and the plots with the current status of the mode identification.

4.6.1   Best models

This table lists the parameters of the 20 best fitting models. The first column always shows the χ_ν^2-value. The other columns contain the parameters that have been set as free for the mode identification. Two display options are possible and can be selected in the combo box above:

- **Best models**
  List the free parameters of the models having the lowest χ_ν^2-values.

- **Best (l,m)-combinations**
  List the free parameters of the models having the lowest χ_ν^2-values *and* different (ℓ,m)-combinations. For each possible combination of ℓ and m, the best model is shown.

4.6.2   Chi-square plots

This plot-window displays the χ_ν^2-values (log-scale) of all models that have been computed in the current optimisation as a function of the free parameters. The uncertainty of the fit for the different parameters can thus be estimated. The free parameter can be selected in the combo box above. By selecting *Model #* in the combo box, the temporal evolution of the χ_ν^2-values during the optimisation is plotted.

4.6.3   Comparison between fit and observation

This box shows the fit of the theoretical values to the observations. The content depends on the selected mode identification method and is described below. The model can be selected by clicking on the corresponding row in the table of best models. The observed values are shown as blue line or symbols, the statistical uncertainty as a green line, and the modelled values as a red line.

- **Compute vsini, EW, intrinsic width, and velocity offset (fit Z)**
  The observed (blue line with uncertainty range as green line) and synthetic (red line) zero-point profile are displayed.

- **FPF methods**
  Three panels are displayed: zero-point (top panel), amplitude (middle panel), and phase (bottom panel) in units of 2π are shown as a function of Doppler velocity (km s^{-1}). The fit is shown as a red line, whereas the observed values are shown as a blue line with the uncertainty range indicated by green lines. The fit for a certain frequency can be selected in the combo box above.

- **Moment method**
  The complete time series of observed and modeled moments is shown. The two panels show the first and second moment, respectively.

4.6.4 List of calculations

This box lists all previously performed optimisations. By clicking on an item in the list, the corresponding parameters are shown in the other windows of this tab.

4.7 Logbook

The logbook provides the list of actions that have been performed with FAMIAS and corresponding information. Each time an operation is carried out in FAMIAS, a new log-entry is written to the *List of actions*. Clicking on an entry of this list shows the corresponding information in the text box.

Entries of the *List of actions* can be renamed or deleted by using the menu *Data*. The text box can be modified and saved in FAMIAS by clicking on the button *Save*.

4.8   Tutorial: Spectroscopic mode identification

This tutorial demonstrates how to perform a mode identification of a time series of synthetic spectra with FAMIAS. The synthetic spectra can be found in the installation directory of FAMIAS in the directory tutorial/*.coasttutorial. The synthetic data simulate spectroscopic observations of a multi-periodic δ Scuti star which consist of one absorption line having realistic observation times and signal-to-noise ratio. The time-series contains 490 spectra that consist of 91 and 77 pixels, respectively, and cover a wavelength range between 5381 and 5385 Ångstrom. These spectra have been computed using the tool *Line Profile Synthesis* of FAMIAS. The input parameters of the model can be found in the file tutorial/coasttutorial.star.

References to functions of FAMIAS are written in the following manner: *Main Window → File → Import Set of Spectra*, which could be translated as: Select in the *Main Window* the function *Import Set of Spectra* in the menu *File*. In each tab, there are named boxes, which can also be referred to. For instance, *Fourier Tab → Settings → Calculations based on: 1st moment* implies that you have to select the *Fourier Tab* and choose the option *1st moment* in the combo box denominated *Calculations based on* in the box called *Settings*.

4.8.1   Import spectra

Follow the following procedure to import the spectra to FAMIAS.

1. Import the spectra by selecting *Main Window → File → Import Set of Spectra*. In the file manager that opens, select the directory tutorial located in the installation directory of FAMIAS and double-click on the file times.coasttutorial. This file contains the observation times and file names of all spectra of this time series. Figure 12 shows a screenshot of FAMIAS after importing the tutorial time series of spectra.

2. The *Import file dialogue* that opens shows the contents of this file. Click *OK* to import this file. In the following dialogue that opens, select Ångstrom as dispersion scale and click *OK*.

3. After successful import, the spectra are displayed as data set in the *Data Manager Tab*. Click into the *Time series* list to display specific spectra in the plot window.

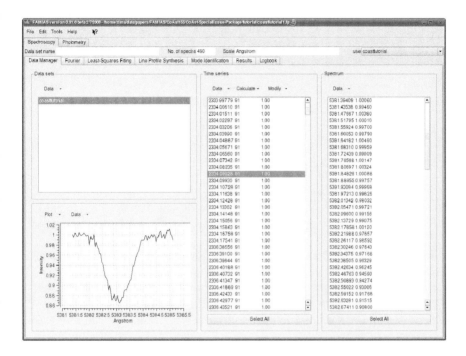

Figure 12: *Data Manager* after importing the tutorial time series of spectra.

4.8.2 Select dispersion range

The synthetic data consist of one absorption line. In general, one has to select a suited spectral line for analysis. Such a line should be an unblended metallic line. Balmer and He-lines are not well suited for the mode identification, since they cannot be well approximated with an intrinsic Gaussian line profile. To study a specific line with FAMIAS, follow the following procedure: Click on one spectrum of the time series. Zoom in on the line in the plot window. Click on *Time Series → Select All* and then select *Time series → Data → Extract Dispersion Range*. You can modify the dispersion range in the dialogue window that opens. After you clicked *OK*, a new data set has been written that only contains the selected dispersion range.

4.8.3 Convert from wavelength to Doppler velocity

To compute moments and to carry out a mode identification, the dispersion scale of the spectra has to be converted from Ångstrom to km s^{-1}. To do so, select all spectra, click on *Time series → Modify → Convert Dispersion* and enter the value of the central wavelength in the dialogue window, which is 5383.369 Ångstrom in this case. The converted spectra are written as a new data set.

4.8.4 Compute signal-to-noise ratio and weights

Computing the SNR of the spectra is important for weighting the spectra, for calculating the statistical uncertainty of the moments, and for enabling the calculation of chi-square with the moment method. There are two ways to estimate the SNR of your spectra.

The simple way is to compute the mean SNR of all spectra. To do so, compute the standard deviation of your spectra (see Section 4.8.7). The mean SNR is the inverse value of the standard deviation at a dispersion position of the continuum.

A more sophisticated and better approach is of course to calculate the SNR of each spectrum separately. If these values have been determined with an external program, they can be imported with the list of times and file names. The weight of each spectrum can then be computed with the function *Time series → Calculate → Compute weights from SNR*. To compute the SNR with FAMIAS, select the function *Time series → Calculate → Compute Signal-to-Noise Ratio*. Adapt the parameters *Factor for sigma clipping* and *Number of iterations* in such a way that only continuum is selected in all spectra. Click on *Write signal-to-noise ratio as normalised weights* to write a new weighted data set.

4.8.5 Compute moments

1. Compute the SNR and weights as described in Section 4.8.4. Before computing the moments, the spectral line should be extracted by excluding the continuum. We refer to Section 4.1.2 for a detailed description how to extract a spectral line with FAMIAS. If the line borders do not move significantly due to the pulsation (=low radial velocity), one can cut out the line, assuming fixed left and right limits. The position of these limits can be determined by interpolating the dispersion scale of the spectra onto a common scale (see Section 4.8.6), computing the mean spectrum with *Time series → Calculate → Mean Spectrum*, and noting the Doppler velocity of the left and right borders of the line (transition to the continuum). Select the original data set (non-interpolated) and extract the line with *Time series → Data → Extract dispersion range*. The extracted spectra are written into a new data set. Select this data set and press *Select All*.

2. Call the dialogue for computing the moments by pressing *Time series* → *Calculate* → *Compute Moments*. Select *Individual signal-to-noise ratio* and the moment that you want to compute in the combo box below. Press *OK* and leave the centroid velocity at the proposed value (=mean barycentre of the line).

3. The time series of moments is written as a new data set. It is advisable to check for systematic trends, especially in the equivalent width and the first moment.

4.8.6 Interpolate on common dispersion scale

The tutorial spectra have different dispersion scales. Therefore, they have to be interpolated onto a common dispersion scale to carry out several tasks, such as to compute a two-dimensional Fourier transform or a least-squares fit across the line profile (pixel-by-pixel), and to apply the FPF method. This is not mandatory when only line moments are used. To carry out a linear interpolation on a common dispersion scale, select all spectra and click on *Time series* → *Modify* → *Interpolate Dispersion*. It is advisable to interpolate onto the spectrum having the highest resolution in order not to lose information. In our case, the first spectrum of the time series has the highest resolution. Therefore, select the function *Interpolate onto scale of first spectrum* and click *OK*. The following dialogue window shows the dispersion values of the mask. To start the interpolation, click on *OK*.

4.8.7 Compute line statistics

The temporal weighted mean of the spectra can be computed with the function *Time series* → *Calculate* → *Mean Spectrum*. To check for line profile variability and estimate the SNR, the standard deviation at each pixel of the spectrum can be computed with the function *Time series* → *Calculate* → *Std. Deviation Spectrum*.

4.8.8 Searching for periodicities

It is advisable to search for periodicities in the data in the pixels across the line profile as well as in the line moments. For the first approach, a data set should have the following properties: interpolated on a common dispersion scale, converted to km s^{-1}, and weighted. For the analysis of the line moments, the data should be converted to km s^{-1}, weighted, and the SNR should be computed for each spectrum. The two approaches can reveal pulsation modes having different characteristics.

The search for periodicities should be carried out in the following iterative schematic way:

1. Compute a Fourier spectrum in a frequency range where you expect pulsation.

2. Compute the significance level at the frequency having the highest amplitude and include this frequency in the least-squares fitting if it is significant.

3. Compute a multi-periodic least-squares fit of the original data with all detected frequencies. In case that no unique frequency solution exists due to aliasing, compute least-squares fits for different possible frequency sets. The solution resulting in the smallest residuals should be regarded as best solution.

4. Exclude frequencies from the fit that do not have a SNR above 4 (3.5 for harmonics/combination terms).

5. Pre-whiten the data with all significant frequencies.

6. Continue with the first point using the pre-whitened data until no significant frequency can be found.

Line moments

1. **Select data set**
   Select the spectra that were prepared for the analysis of the line moments and go to the *Fourier Tab*.

2. **Calculate Fourier spectrum of equivalent width**
   Select the option *Fourier Tab → Settings → Calculations based on → Equivalent width* and click on *Calculate Fourier*. The plot window now displays the Fourier spectrum of the equivalent line width. A dialogue opens, indicating the highest frequency peak at $F_1 = 3.148$ d^{-1} and asking if this frequency should be added to the frequency list of the *Least-Squares Fitting Tab*. Since we first want to check the significance of this frequency, click on *No*.

3. **Compute significance level**
   Select the option *Settings → Compute significance level*. The field *at frequency* should now contain the value 3.148038. Compute the Fourier spectrum once more by clicking on *Calculate Fourier*. The plot window now also displays the significance level as a red curve and the dialogue window indicates the SNR of the highest peak. Since it has a SNR of 4.1, click on *Yes* to include it in the frequency list.

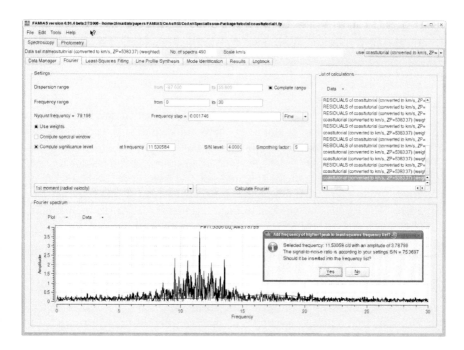

Figure 13: Fourier spectrum of the first moment after pre-whitening with $F_2 = 11.53 \text{ d}^{-1}$.

4. Compute least-squares fit

Go to the *Least-Squares Fitting Tab* and select the field *Settings* →
Compute signal-to-noise ratio and the frequency F1. Compute a least-squares fit by pressing *Settings* → *Calculate Amplitude + Phase*. Improve the frequency solution by clicking *Settings* → *Calculate All*. According to the *List of frequencies*, this frequency has a SNR of 3.96, which is just below the significance limit. The difference with the SNR determined in the Fourier transform is due to the fact that, in this case, the amplitude determined from the least-squares fit is taken as signal. We can conclude now that there are no significant periodic equivalent width variations in the line profile.

5. Calculate Fourier spectrum of first moment

Select the option *Fourier Tab* → *Settings* → *Calculations based on* →
1st moment and click on *Calculate Fourier*. The highest peak is at the frequency $F_2 = 11.53 \text{ d}^{-1}$. Check for significance as described in the previous point. Since this peak is highly significant, it should be included

in the *List of frequencies*. A screenshot of FAMIAS showing the Fourier spectrum of the first moment is displayed in Figure 13.

6. **Compute least-squares fit and pre-whiten data**
Select the detected frequency F_2 in the *Least-Squares Fitting Tab*, compute a least-squares fit and pre-whiten the data (*Settings → Pre-whiten data*). The residuals are written as a new data set in the *Data Manager Tab*. The *List of frequencies* shows the results of the computed fit and the derived uncertainties of the parameters. The value of the field *Residuals* is computed from the standard deviation of the residuals. The frequency is always indicated in units of the inverse of the input timestring. The units of the amplitude depend on the selected calculation basis. The equivalent width is in units of km s^{-1}. The n-th moments is in units of $(\text{km s}^{-1})^n$. The phase is in units of 2π.

7. In the *Data Manager Tab*, select the time series of residuals and check the computed fit (red line).

8. If you want to compute a Fourier spectrum or a least-squares fit of line moments, you have two possibilities. The first option is to compute the moments of the line profile in the *Data Manager Tab* (see Section 4.8.5) and then analyse this one-dimensional time series. The other possibility is to choose a time series of spectra and then to select the option *Settings → Calculations based on → n-th moment* in the *Fourier Tab* and the *Least-Squares Fitting Tab*. In this case, the corresponding time-series of moments is computed automatically for the selected dispersion range. If you then pre-whiten your data, the residuals are written as time series of moments to the *Data Manager Tab* (yellow background in the list of *Data sets*. You can calculate a Fourier transform of these residuals to search for further peaks. If you want to compute another least-squares fit with an additional frequency, you have to select the original time-series of spectra (green background in the list of *Data sets*).

9. **Compute Fourier spectrum of residuals**
Compute a Fourier spectrum of the residuals. A frequency at $F_3 = 17.5$ d^{-1} is significant and should also be included in the least-squares fit. Also select this frequency in the *Least-Squares Fitting Tab → List of frequencies* and compute a least-squares fit for both frequencies simultaneously. You have to select the original data set that was prepared for computing the moments. Pre-whiten the data and compute a Fourier spectrum of the residuals. No significant peaks are left.

10. **Analyse the first three moments**
Analyse also the second and third moments since modes of higher degree

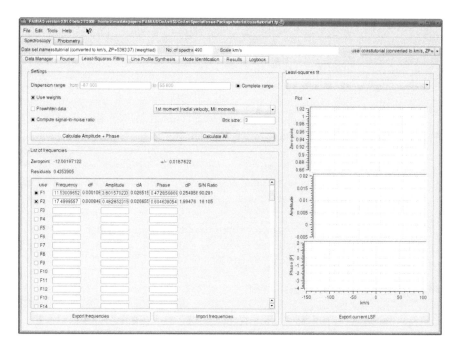

Figure 14: Results of the least-squares fit to the first moment.

might only have significant amplitudes for these diagnostics. The analysis of the second moment should reveal F_2, $2F_2$, $2F_3$, and an additional frequency at $F_4 = 23.998\ \mathrm{d}^{-1}$. The third moment only has F_2 as significant peak.

11. We can conclude that three significant independent frequencies are present in the first three moments of the tutorial data. Only two of them are visible in the first moment and thus analysable with the moment method. Figure 14 displays a screenshot of FAMIAS showing the results of the least-squares fit to the first moment.

Pixel-by-pixel across the line profile

1. **Select data set**
   Select the data set that was prepared for the frequency analysis across the line profile (pixel-by-pixel) and go to the *Fourier Tab*.

2. **Calculate Fourier spectrum**
   Select the option *Fourier Tab* → *Settings* → *Calculations based on*

$\rightarrow$ *Pixel-by-pixel (1D, mean Fourier spectrum)* and click on *Calculate Fourier.*

3. **Compute significance level**
   The plot window now shows the mean of all Fourier spectra across the line profile. A dialogue opens, indicating the highest frequency peak at $F_1 = 11.53$ d^{-1}. Since we first want to check the significance of this frequency, click on *No.* To determine the significance of F_1, check the field *Settings* $\rightarrow$ *Compute significance level* and select the option *Settings* $\rightarrow$ *Calculations based on* $\rightarrow$ *Pixel with highest amplitude at* f=. The latter option is necessary, since the significance level cannot be determined from the mean Fourier spectrum across the line profile. Click *Calculate Fourier* to compute the Fourier spectrum and its significance level at the dispersion position, where the given frequency has its highest amplitude. Since this frequency is highly significant, add it to the *List of frequencies* in the *Least-Squares Fitting Tab.*

4. **Compute least-squares fit**
   In the *Least-Squares Fitting Tab*, select the option *Settings* $\rightarrow$ *Calculations based on* $\rightarrow$ *Pixel-by-pixel (MI:FPF)*, check the box next to the detected frequency, and press *Calculate Amplitude + Phase* to compute the least-squares fit. Zero-point, amplitude, and phase will be displayed in the plot panel at the right-hand side. The blue lines denote the derived fit, whereas the green lines indicate the statistical uncertainty range of the fit.

5. The *List of frequencies* shows the results of the computed fit. The field *Results* shows the mean standard deviation of the residual spectra. The frequency is indicated in inverse units of the input time-string. The *IAD* is the integrated amplitude distribution, and is calculated from the integral of the amplitude across the line profile inside the selected dispersion range.

6. **Pre-whiten spectra**
   Pre-whiten the data with the determined least-squares fit by checking the box *Settings* $\rightarrow$ *Pre-whiten data* and clicking *Calculate Amplitude + Phase.* The pre-whitened spectra are written as a new data set to the *Data Manager Tab.*

7. Select the time series of residual spectra in the *Data Manager Tab* and click on several spectra to check the quality of the fit (red line).

8. **Compute Fourier spectrum of residuals**
   Compute a Fourier spectrum of the residuals by selecting the time series

of residual spectra in the *Data Manager Tab* and proceeding as described in point 2.

9. When computing further multi-periodic least-squares fits, the original time series of spectra has to be selected.

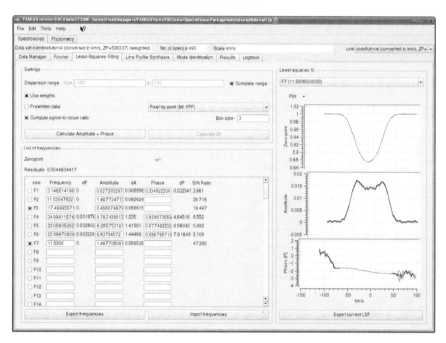

Figure 15: Results of the least-squares fit across the line profile.

10. The period analysis of the tutorial data set reveals three frequencies, $F_1 = 11.53 \ \mathrm{d}^{-1}$, $F_2 = 17.50 \ \mathrm{d}^{-1}$, and $F_3 = 2F_1 = 23.06 \ \mathrm{d}^{-1}$. The frequency F_3 is a harmonic of F_1.

4.8.9 Mode identification

FAMIAS provides two different approaches for the spectroscopic mode identification, the moment method and the Fourier parameter fit method. In the following, we will describe in detail the approach for each method separately.

- **Setting the parameters**
  Parameters on the *Mode Identification Tab* that have a check box next to the parameter name can be set as variable during the optimisation. In

this case, a minimum, maximum, and step value have to be indicated. If the box is unchecked, the parameter is set as constant during the optimisation with the value indicated.

- **Stellar parameters**
  You need to provide estimates for the stellar radius, mass, T_{eff}, $\log g$, and metallicity in the field *Stellar Parameters*. The indicated radius and mass mainly affect the numerical calculation of the horizontal to vertical pulsation amplitude and can be set as variable during the optimisation. The three other parameters determine the limb darkening coefficient, which is interpolated linearly in a pre-computed grid (Claret et al. 2000). The inclination and $v \sin i$ can be fixed, when they are known. Otherwise, they can be estimated during the mode identification and should be set as variable in a reasonably large range (see Figure 16).

- **Line Profile Parameters**
  The only parameter which has to be known a priori is the *Central wavelength* of the considered line profile. This parameter determines the adopted limb darkening coefficient. If one deals with a cross correlated profile, this value of course does not have a physical meaning. In this case, it is best to enter the mean value of the cross correlated range into this field. The other parameters in this field can be determined during the mode identification.

  In the case of the moment method, also the centroid velocity and the signal-to-noise ratio have to be set. See Section 4.5.9 for details about these parameters.

- **Pulsation Mode Parameters**
  This field controls the settings for the parameters of each imported pulsation frequency. A frequency will be taken into account for the optimisation if the check box next to the field *Frequency [c/d]* is checked.

- **Optimisation Settings**
  This field controls how the mode identification is applied as well as the settings for the genetic optimisation. For a detailed explanation of the settings, we refer to Section 4.5.10.

- **General Settings**
  For the tutorial spectra, you can leave the number of segments that are taken into account for the computation of the line profile at the value of 1000. For rapidly rotating stars and high-degree pulsation modes, a higher value is required. For details, we refer to Section 4.5.11. FAMIAS

proposes default parameter settings for the optimisation if you press *Set fields to default.*

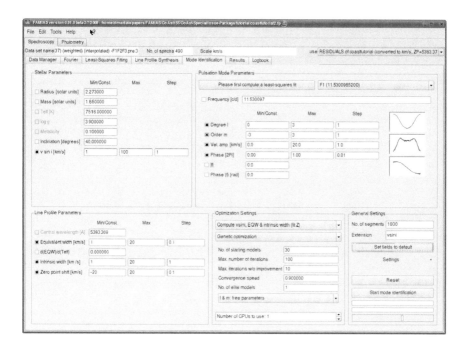

Figure 16: Settings of the *Mode Identification Tab.*

Fourier parameter fit method

1. **Determine pulsation frequencies**

   Determine all pulsation frequencies, including harmonics and combinations, that have significant amplitude across the line profile (=pixel-by-pixel) as described in the previous section. Select all frequencies in the *List of frequencies* and compute a least-squares fit across the line profile with the option *Settings* → *Calculations based on* → *Pixel-by-pixel (MI:FPF).*

2. **Selection of dispersion range**

   After you have imported the current least-squares fit to the mode identification tab, the dispersion range that is taken into account for the mode identification can no longer be modified. Therefore, you have to define the dispersion range already when you compute the least-squares fit. An optimal range excludes the continuum and the line wings. Only the range

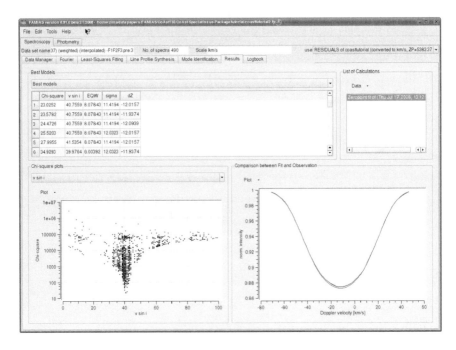

Figure 17: Results of the fit to the zero-point profile.

where the amplitude across the profile reaches significant values should be selected. You can either modify the dispersion range in the field *Settings* or zoom into the selected region in the plot window. In the latter case, the left and right dispersion values of the zoomed range are automatically written to the *Settings*-field. Uncheck the box *Settings* → *Complete range* and compute a least-squares fit. In the tutorial example a range between -70 and 45 km s^{-1} would be optimal.

3. **Import frequencies to Mode Identification Tab**
   Switch to the *Mode Identification Tab* and import the current multi-periodic least-squares fit by clicking on *Pulsation Mode Parameters* → *Import data for FPF method (from current LSF)*. In the field *Pulsation Mode Parameters* you can now switch between the different imported pulsation frequencies.

4. **Determine pulsationally independent parameters**
   For the tutorial spectra, the stellar parameters have been saved in a file called coasttutorial.star. You can import this file by selecting *General Settings* → *Settings* → *Import stellar parameters*. We will first de-

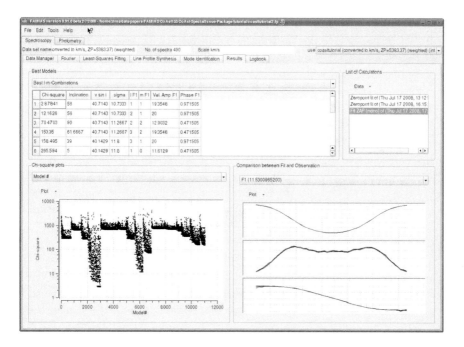

Figure 18: Results of the mode identification for $F_1 = 11.53$ d^{-1}.

termine starting values for the pulsationally independent parameters, i.e., $v \sin i$, the equivalent width, the intrinsic width, and the velocity zero-point shift of the profile. The search range of these parameters should be sufficiently large with a reasonable step-size. For the tutorial example, good starting values would be [min;max;step]: $v \sin i \in [1; 100; 1]$, equivalent width $\in [1; 20; 0.1]$, intrinsic width $\in [1; 20; 1]$, and zero-point shift $\in [-20; 20; 0.1]$. The step width should generally not be smaller than the precision to which a parameter can be determined. The best approach is to begin with a relatively large search range and step size, and to iteratively narrow the range. See Figure 16 for a screenshot of the *Mode Identification Tab* with the settings before the first optimisation.

Select the option *Select MI method → Compute vsini, EW, intrinsic width, and velocity offset (fit Z)* and press on *General Settings → Set fields to default* to set default parameters for the genetic optimisation and to let FAMIAS propose free parameters for the optimisation. In this case, $v \sin i$, the equivalent width, the intrinsic width, and the zero-point shift are set as free.

Press *General Settings* → *Start mode identification* to start the optimisation. The results are written to the *Results Tab*. Figure 17 shows the results of the fit to the zero-point profile. The field *Best Models* shows the 20 best solutions. You can click into the table to display the fit in the plotting window *Comparison between Fit and Observation*. Here, the observational zero-point profile is displayed as blue line, its statistical uncertainty as green lines, and the modeled profile as red line. The chi-square plots can be used to estimate the uncertainty of the fit.

It is evident in Figure 17 that the best solution can still be improved. It is a good idea to note the parameter values for the best solutions and refine the search range of these parameters in the *Mode Identification Tab* to the following values [min;max;step]: $v \sin i \in [30; 50; 1]$, equivalent width $\in [7; 9; 0.01]$, intrinsic width $\in [7; 15; 1]$, and zero-point shift $\in [-13; -11; 0.01]$. Reset the optimisation procedure by pressing *General Settings* → *Reset* and start another optimisation.

This optimisation should result in a much lower chi-square value and thus a better constraint on the free parameters. We will take the obtained solution as a starting point for the mode identification. Generally, the equivalent width and the zero-point shift are quite well constrained and can be set as constant during the optimisation. Figure 18 shows the parameter range we selected for the mode identification.

5. **Identify pulsation modes**
   Select the Fourier parameter fit method with the option to fit zero-point, amplitude, and phase across the line profile through the combo box *Optimisation Settings* → *Select MI method* → *FPF Method: fit ZAP*. Select the option *Optimisation Settings* → *l & m: grid search* to obtain more reliable results of the optimisation procedure. Click on *General Settings* → *Set fields to default*. The inclination i is now also set as free parameter. Enter the following values as range: $i \in [5; 90; 10]$. Select the frequency 11.53 d^{-1} in the field *Pulsation Mode Parameters* and mark the check box next to the frequency value. The parameter ranges should be as follows: degree $\ell \in [0; 3; 1]$, order $m \in [-3; 3; 1]$, vel.amp $v \in [0; 30; 1]$, and phase $\psi \in [0.4715; 0.9715; 0.5]$. The value of the phase is taken from the least-squares solution of the first moment of this frequency (see Section 4.5.4 for details). The maximum value of the velocity amplitude should be set at least an order of magnitude higher than the amplitude of the first moment. In general, you should extend the range of a parameter, if the lowest chi-square value was found at one of the search border (minimum or maximum of the range). Start the optimisation by pressing *Start mode identification*. The other pulsation mode at

17.5 d^{-1} can be analysed in the same manner. You can compare your results with the input values by loading the file coasttutorial.star into the *Line Profile Synthesis Tab.*

All computed mode identifications are saved in the *Results Tab* → *List of Calculations* and are logged in the *Logbook* of FAMIAS.

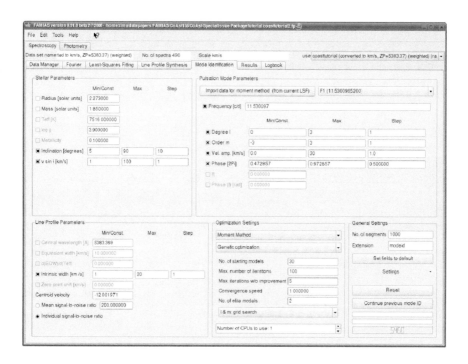

Figure 19: Settings of the *Mode Identification Tab* for the moment method.

Moment method

1. **Determine pulsation frequencies**
   Determine all frequencies that have significant amplitude in the first moment, including harmonics and combination frequencies (see previous section). Compute a multi-periodic least-squares fit using the option *Least-Squares Fitting Tab* → *Settings* → *Calculations based on* → *1st moment (radial velocity, MI: moment).*

2. **Import frequencies to Mode Identification Tab**
   Switch to the *Mode Identification Tab* and click on *Pulsation Mode Parameters* → *Import data for moment method (from current LSF).*

After the import, you can switch between the two pulsation frequencies in the field *Pulsation Mode Parameters* with the top left combo box. Mark the check box next to the frequency value for both imported frequencies.

3. **Identify pulsation modes**
   The starting parameters and settings should be adopted as displayed in Figure 19. Start the mode identification by clicking on *General Settings* → *Start mode identification*. The results are written to the *Results Tab*.

5. The Photometry Modules

The *Photometry Module* contains tools that are required to search for frequencies in photometric time series and to perform a photometric mode identification. The tools are located in tabs that have the following denominations: *Data Manager, Fourier, Least-Squares Fitting, Mode Identification, Results,* and *Logbook*. These tools are described in the following sections.

5.1 Data Manager

The *Data Manager Tab* gives information about light curves that have been imported, allows to edit them, and select the data sets for analysis. The window is divided into two data boxes and one plot window. A menu is located above each box. In the *Data Sets Box* you can select the light curve you want to analyse. The *Time Series Box* displays the time of measurement, magnitude, and weight of the selected data set. The *Plot Window* displays the currently selected light curve and data points that have been selected in the *Time Series Box*. A screenshot of the *Data Manager* is displayed in Figure 20.

5.1.1 Data Sets Box

This box contains a list of the different data sets that have been imported or created. To select a data set, click on it or select it in the combo box at the top right of the information bar. The following commands can be selected in the *Data Menu*:

- **Remove Data Set**
  Removes the currently selected data set from the list.

- **Rename Data Set**
  Renames the currently selected data set.

- **Export Data Set**
  Exports the currently selected light curve as an ASCII-file to the disk. The suffix of the files has to be entered by the user. The exported files have the following three columns: time, magnitude, and weights.

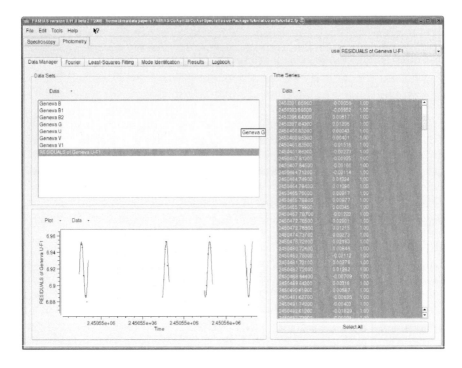

Figure 20: Screenshot of the *Data Manager Tab.*

- **Combine Data Sets**
  Combines the selected data sets to a new single time series. The data sets
  to be combined must have the same units of the dispersion. Moreover,
  all times of measurement have to differ.

- **Change Assigned Filter**
  Change the filter that is assigned to the current time string. The correct
  filter has to be assigned to assure correct working of the mode identification.

5.1.2 Time Series Box

This list shows the measurements of the currently selected light curve. It consists of three columns: times of measurement, magnitude, and weight. Multiple
measurements can be selected by clicking with the left mouse button on several
items in the list while pressing the Ctrl-key or the Shift-key. All items can be
selected by pressing *Select All. Only items that have been selected in this list*

(with blue background) are taken into account for the data analysis (e.g., Fourier analysis or least-squares fitting). Selected items are marked with a red cross in the *Plot Window*.
The following commands are available in the *Data Menu*:

- **Edit Data**
  Opens a table of times and weights in a new window with the possibility to edit these values. Modifications can be written to the current data set.

- **Copy Selection to New Set**
  A new data set with currently selected measurements is created and written to the *Data Sets Box*. Use this option to create subsets of your data.

- **Remove Selection**
  The currently selected measurements are removed from the time series/data set.

5.1.3   Plot window

The plot window shows the currently selected light curve as blue symbols. Selected measurements are marked with a red cross.
For more information about the plot window, we refer to p. 28.

5.2   Fourier Analysis

With this module, a discrete Fourier transform (DFT) can be computed to search for periodicities in the data set selected in the *Data Sets Box* of the *Data Manager Tab*. The Fourier spectrum is displayed in the plot window and saved as data set in the *List of calculations*. A screenshot of the *Fourier Tab* is displayed in Figure 21.

5.2.1   Settings Box

In this box, the settings for the Fourier analysis are defined.

- **Frequency range**
  Minimum/Maximum values of the frequency range. The Fourier spectrum will be computed from the minimum to the maximum value.

- **Nyquist frequency**
  Estimation of the Nyquist frequency (mean sampling frequency). For non-equidistant time series, a Nyquist frequency is not uniquely defined. In this case, the Nyquist frequency is approximated by the inverse mean of the time-difference of consecutive measurements by neglecting large gaps.

- **Frequency step**
  Step size (resolution) of the Fourier spectrum. Three presets are available: Fine ($\equiv (20\Delta T)^{-1}$), Medium ($\equiv (10\Delta T)^{-1}$), and Coarse ($\equiv (5\Delta T)^{-1}$). The corresponding step size depends on the temporal distribution of the measurements, i.e., the time difference ΔT of the last and first measurement. It is recommended to select the fine step size to ensure that no frequency is missed. The step value can be edited if desired.

- **Use weights**
  If the box is checked, the weight indicated for each data point is taken into account in the Fourier computations. Otherwise, all weights are assumed to have equal values.

- **Compute spectral window**
  If the box is checked, a spectral window of the current data set is computed. A spectral window shows the effects of the sampling of the data on the Fourier analysis and thus permits to estimate aliasing effects. The spectral window is computed from a Fourier spectrum of the data taking the times of measurements and setting all intensities to the value 1. The shape of the spectral window should be plotted for a frequency range that is symmetric around 0 for visual inspection.

- **Compute significance level**
  If the box is checked, the significance level at a certain frequency value is computed and displayed in the plot window as a red line. The following parameters can be set:

  - **Frequency**
    Frequency value of the peak of interest. The data will be pre-whitened with this frequency and the noise level will be computed from the pre-whitened Fourier spectrum.

  - **S/N level**
    Multiplicity factor of the signal-to-noise level. The displayed noise level will be multiplied by this factor.

  - **Box size**
    Box size b for the computation of the noise-level in units of the frequency. The significance level is computed from the running mean of the pre-whitened Fourier spectrum. For each frequency value F, the noise level is calculated from the mean of the range $[F-b/2, F+b/2]$.

- **Calculate Fourier**
  Computes the discrete Fourier transform (DFT) according the user's settings and displays it in the plot window as a blue line. The mean intensity value of the time series is automatically shifted to zero before the Fourier analysis is computed. The peak having highest amplitude in the given range is marked in the plot window. A dialogue window reports the frequency having the highest amplitude in the selected frequency range and asks if it should be added to the frequency list of the *Least-Squares Fitting Tab*.

5.2.2 List of Calculations

Previous Fourier calculations can be selected from the list and viewed. Each computed Fourier spectrum is saved and listed here. If a project is saved, the list of computed Fourier spectra is also saved but compressed to decrease the project file size (only extrema are saved). The following operations are possible via the *Data Menu*:

- **Remove Data Set**
  Removes the currently selected data set from the list.

- **Rename Data Set**
  Renames the currently selected data set.

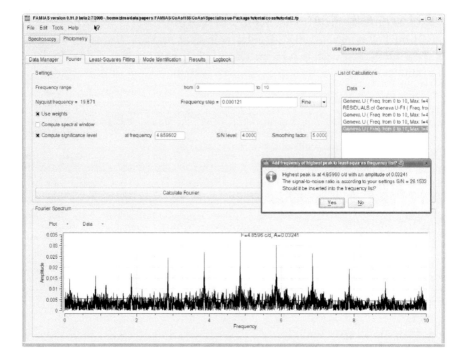

Figure 21: Screenshot of the *Fourier Tab*.

- **Export Data Set**
  Exports the currently selected data set to an ASCII file having the follow-ing three-column format: frequency, amplitude, power.

5.2.3   Fourier Spectrum Plot

Shows the most recently computed Fourier analysis or the selection from the list of calculations. The Fourier spectrum is shown as a blue line, the significance level is shown as a red line. The frequency and amplitude of the peak having the highest frequency are indicated.

For more information about the plot window, we refer to p. 28.

5.3   Least-Squares Fitting

This modules provides tools for the computation of a non-linear multi-periodic least-squares fit of a sum of sinusoidals to your data. The fitting formula is

$$Z + \sum_i A_i \sin\left[2\pi(F_i t + \phi_i)\right] \tag{22}$$

Here, Z is the zero-point, and A_i, F_i, and ϕ_i are amplitude, frequency and phase (in units of 2π) of the i-th frequency, respectively.

The least-squares fit is carried out with the Levenberg-Marquardt algorithm. For a given set of frequencies, either their zero-point, amplitude and phase can be optimized (*Calculate Amplitude & Phase*), or additionally also the frequency value itself (*Calculate All*). The data can be pre-whitened with the computed fit and written to the *Data Sets Box* of the *Data Manager Tab*.

Before a mode identification can be carried out, a least-squares fit to the data must be calculated. To carry out a photometric mode identification, light curves from different filters must be imported to FAMIAS, and amplitudes and phases of the pulsation frequencies must be determined by least-squares fitting. These values can then be copied to the *Mode Identification Tab* to carry out the mode identification method using amplitude ratios and phase differences.

5.3.1   Settings

Defines the settings for the calculation of the least-squares fit.

- **Use weights**
  If this box is checked, the weight indicated for each data point is taken into account in the least-squares fit. Otherwise, all weights are assumed to have equal values.

- **Pre-whiten data**
  If this box is checked, the data will be pre-whitened with the computed least-squares fit and written as a new data set to the *Data Manager Tab*.

- **Compute signal-to-noise ratio**
  Computes the amplitude signal-to-noise ratio (SNR) of each selected frequency and displays it in the *List of Frequencies*. The noise is computed from the Fourier spectrum of the pre-whitened data. The *Box size* is the width of the frequency range which is taken into account for the calculation of the noise. For a box width of b, the noise of frequency F is the mean value of the Fourier spectrum of the residuals in the range $[F - b/2, F + b/2]$. The SNR is the ratio of A_f and the noise level of the pre-whitened Fourier spectrum at the position of f.

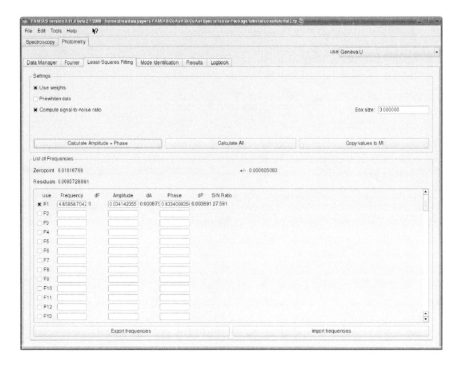

Figure 22: Screenshot of the *Least-Squares Fitting Tab.*

- **Calculate Amplitude + Phase**

  Computes a least-squares fit with the Levenberg-Marquardt algorithm using the above mentioned fitting formula. The zero-point, amplitude and phase are calculated, whereas the frequency is kept fixed.

  The following optimized values are written to the frequency list: zero-point and its uncertainty, the standard deviation of the residuals, for each selected frequency its amplitude and phase and their formal uncertainties derived from the error matrix of the least-squares fitting algorithm.

- **Calculate All**

  Computes a least-squares fit with the Levenberg-Marquardt algorithm using the above mentioned fitting formula. The zero-point, amplitude, phase and frequency are improved. The following optimized values are written to the frequency list: zero-point and its uncertainty, the standard deviation of the residuals, for each selected frequency its frequency value, amplitude and phase and their formal uncertainties derived from the error matrix of the least-squares fitting algorithm.

- **Copy values to MI**
  Computes a least-squares fit by improving amplitude and phase (equivalent to *Calculate Amplitude + Phase*) and copies the derived values (frequencies, amplitudes, and phases and their uncertainties) to the *Mode Idenitification Tab*. For different filters, a least-squares solution with identical frequency values must be computed to ensure that the phases in the different filters can be compared.

5.3.2 List of Frequencies

The *List of Frequencies Box* shows the results of the least-squares fit. Frequencies that should be included in a least-squares fit can be entered in the column *Frequency* and selected by clicking on the check box in column *Use*. The following values are shown in this box after a least-squares fit has been calculated:

The zero-point, its formal uncertainty and the standard deviation of the residuals are shown at the top. The improved values of frequency, amplitude and phase and their formal statistical uncertainties are shown in the list. The phase and its uncertainty, in units of 2π. The last column lists the SNR of each frequency (only shown when option Calculate signal-to-noise ratio has been checked). The SNR is computed from the Fourier spectrum, pre-whitened with all selected frequencies. For each frequency, the assumed noise-level is computed from the mean amplitude around the frequency value with the box size indicated at the option Calculate signal-to-noise ratio.

- **Export frequencies**
  Exports all frequency, amplitude and phase values of the List of frequencies to an ASCII file. The file format is compatible with the program Period04 (see example on p. 48).

- **Import frequencies**
  Imports an ASCII list of frequencies having the following four-column format separated with tabulators: frequency counter, frequency value, amplitude, phase (see example on p. 48).

5.4   Mode Identification

This module can be used to perform a photometric mode identification based on the method of amplitude ratios and phase differences of pulsation modes in different photometric passbands (Balona & Stobie 1979; Watson 1988; Cugier et al. 1994). This method permits to determine the harmonic degree ℓ of pulsation modes in general up to $\ell = 6$. This upper limit is due to partial geometric cancelation of the observable pulsation amplitude over the stellar disc.

The determination of the ℓ-degrees is based on static plane-parallel models of stellar atmospheres and on linear non-adiabatic computations of stellar pulsation. In the present version of FAMIAS, these are provided in the form of pre-computed grids and interpolated linearly to obtain values appropriate for the observed parameters. The theoretical values of the amplitude ratio and phase difference in a certain filter depend strongly on pulsational input. This points out a very important difference between spectroscopic and photometric mode identification: the former is model independent, the latter is not. To be able to compare the results, FAMIAS incorporates grids computed from different pulsational codes and from different atmosphere models. The present version of FAMIAS includes grids from two different scientific institutions (see details below). It is planned to include model grids from more groups in the future, whenever they are provided.

5.4.1   Theoretical background

In FAMIAS, we apply the approach proposed by Daszyńska-Daszkiewicz et al. (2002) to compute the theoretical photometric amplitudes and phases due to pulsation. For more details see instruction on the Wrocław HELAS Webpage[1]. In the limit of linear pulsation, zero-rotation approximation and assuming static plane-parallel atmospheres, we can write the flux variations in the passband λ caused by a oscillation mode having a frequency ω and a degree ℓ as

$$\frac{\Delta\mathcal{F}_\lambda}{\mathcal{F}_\lambda^0} = \varepsilon Y_\ell^m(i,0)b_\ell^\lambda \mathrm{Re}\{[D_{1,\ell}^\lambda f + D_{2,\ell} + D_{3,\ell}^\lambda]e^{-i\omega t}\}, \qquad (23)$$

where

$$D_{1,\ell}^\lambda = \frac{1}{4}\frac{\partial \log(\mathcal{F}_\lambda|b_\ell^\lambda|)}{\partial \log T_{\mathrm{eff}}},$$
$$D_{2,\ell} = (2+\ell)(1-\ell), \qquad (24)$$
$$D_{3,\ell}^\lambda = -\left(\frac{\omega^2 R^3}{GM}+2\right)\frac{\partial \log(\mathcal{F}_\lambda|b_\ell^\lambda|)}{\partial \log g_{\mathrm{eff}}^0}.$$

---

[1]http://helas.astro.uni.wroc.pl/deliverables.php

or equivalently

$$D_{1,\ell}^\lambda = \frac{1}{4}\left(\alpha_{T,\lambda} + \frac{1}{\ln 10}\frac{\beta_{T,\lambda}}{b_\ell^\lambda}\right),$$

$$D_{2,\ell} = (2+\ell)(1-\ell), \tag{25}$$

$$D_{3,\ell}^\lambda = -\left(\frac{\omega^2 R^3}{GM} + 2\right)\left(\alpha_{g,\lambda} + \frac{1}{\ln 10}\frac{\beta_{g,\lambda}}{b_\ell^\lambda}\right).$$

Here, ε is the pulsation mode amplitude expressed as a fraction of the equilibrium radius of the star, $Y_\ell^m(i,0)$ describes the mode visibility with the inclination angle, i, and (ℓ, m) being the spherical harmonic degree and the azimuthal order, respectively, G is the gravitational constant, M is the stellar mass, and b_ℓ^λ is the disc averaging factor defined as

$$b_\ell^\lambda = \int_0^1 h_\lambda^0(\mu)\mu P_\ell(\mu)d\mu. \tag{26}$$

The $D_{1,\ell}^\lambda$ and $D_{3,\ell}^\lambda$ terms describe temperature and gravity effects, respectively, and both include the perturbation of the limb-darkening. The $D_{2,\ell}$ term stands for geometrical effects. For computing b_ℓ^λ, we use a non-linear limb darkening law, defined by Claret et al. (2000) as

$$\frac{I(\mu)}{I(1)} = 1 - \sum_{k=1}^4 a_k(1 - \mu^{\frac{k}{2}}), \tag{27}$$

where $I(\mu)$ is the specific intensity on the stellar disk at a certain line-of-sight angle θ with $\mu = \cos\theta$ and a_k is the k-th limb darkening coefficient.

The parameters $\alpha_{T,\lambda}$ and $\alpha_{g,\lambda}$ are the partial flux derivatives over effective temperature and gravity, respectively, that are calculated from static model atmospheres for different passbands

$$\alpha_{T,\lambda} = \frac{\partial \log F_\lambda}{\partial \log T_{\text{eff}}} \quad \text{and} \quad \alpha_{g,\lambda} = \frac{\partial \log F_\lambda}{\partial \log g}, \tag{28}$$

whereas, the parameters $\beta_{T,\lambda}$ and $\beta_{g,\lambda}$ are partial derivatives of the b_ℓ^λ factor

$$\beta_{T,\lambda} = \frac{\partial \log b_\ell^\lambda}{\partial \log T_{\text{eff}}} \quad \text{and} \quad \beta_{g,\lambda} = \frac{\partial \log b_\ell^\lambda}{\partial \log g}. \tag{29}$$

The f parameter is a complex value which results from linear non-adiabatic computations of stellar pulsation and describes the relative flux perturbation at the level of the photosphere

$$\frac{\delta T_{\text{eff}}}{T_{\text{eff}}^0} = \varepsilon\frac{1}{4}\text{Re}\{fY_\ell^m e^{-i\omega t}\}. \tag{30}$$

According to Eq. (23), the complex amplitude of the light variations is expressed as (Daszyńska-Daszkiewicz et al. 2002)

$$\mathcal{A}_\lambda(i) = -1.086\varepsilon Y_\ell^m(i,0)b_\ell^\lambda(D_{1,\ell}^\lambda f + D_{2,\ell} + D_{3,\ell}^\lambda), \qquad (31)$$

and the amplitudes and phases of the light variation are given by

$$A_\lambda = |\mathcal{A}_\lambda| = \sqrt{\mathcal{A}_{\lambda,R}^2 + \mathcal{A}_{\lambda,I}^2}$$

and

$$\varphi_\lambda = \arg(\mathcal{A}_\lambda) = \arctan(\mathcal{A}_I/\mathcal{A}_R)$$

where

$$\mathcal{A}_{\lambda,R} = -1.086\varepsilon Y_\ell^m(i,0)b_\ell^\lambda(D_{1,\ell}^\lambda f_R + D_{2,\ell} + D_{3,\ell}^\lambda),$$

$$\mathcal{A}_{\lambda,I} = -1.086\varepsilon Y_\ell^m(i,0)b_\ell^\lambda D_{1,\ell}^\lambda f_I.$$

Calculating amplitude ratio and phase differences the $\varepsilon Y_\ell^m(i,0)$ term goes away making these observables independent of the inclination angle, i, and the azimuthal order, m, in the case of zero-rotation approximation.

5.4.2   Approach for mode identification in FAMIAS

FAMIAS computes the theoretical amplitude ratios and phase difference according to the above described scheme in different photometric passbands. To identify the spherical harmonic degree, ℓ, the user must provide its frequency, amplitude and phase in different filters, ranges for T_{eff} and $\log g$, a number of stellar model parameters such as mass and metallicity, and the source of the stellar models. FAMIAS then derives the theoretical values from pre-computed model grids and displays the results of the mode identification on the *Results Tab*.

At the time this manual was written, we had pre-computed grids of the parameters $\alpha_{T,\lambda}$, $\alpha_{g,\lambda}$, $\beta_{T,\lambda}$, $\beta_{g,\lambda}$, b_ℓ^λ at our disposal. These atmospheric parameters have been computed by Leszek Kowalczuk and Jadwiga Daszyńska-Daszkiewicz using Kurucz and NEMO atmospheres. The grids are available for the following photometric systems: Johnson/Cousins $UBVRI$, Strömgren $uvby$, and Geneva, and for different values of metallicity parameter [m/H] and microturbulence velocity, ξ_t. All these results can be found on the Wrocław HELAS Webpage[2].

These model grids contain stellar evolution tracks for different masses computed by the Warsaw-New Jersey code (Paczyński 1969, 1970) and pulsational

---

[2]http://helas.astro.uni.wroc.pl/deliverables.php

models from ZAMS to TAMS with some step in time (or effective temperature) computed for mode degree ℓ from 0 to 6. The full description of the evolutionary and pulsational models is given at the Wrocław HELAS Webpage.

Furthermore, we used pulsational models from two different sources available. First, a grid for main-sequence stars with masses from 1.8 to 12 $M_\odot$ computed by Jadwiga Daszyńska-Daszkiewicz, Alosha Pamyatnykh, and Tomasz Zdravkov using the non-adiabatic Dziembowski code (Dziembowski 1971, 1977), which can be downloaded also from the above mentioned web site. Second, a grid for δ Sct stars computed with ATON (Ventura et al. 2007) and MAD by Montalban & Dupret (2007).

The grids included in the present version of FAMIAS cover the following range:

- $1.6 \leq M_\odot \leq 12$

- $3500 \leq T_{\mathrm{eff}} \leq 47500$ K

- $1 \leq \log g \leq 5$

- $-5 \leq [m/H] \leq 1$

- $0 \leq v_{\mathrm{micro}} \leq 8$ km s^{-1} (for some metallicities).

In more detail, the mode identification is carried out in the following way:

- The user must provide the pulsation frequency F, its amplitude A_λ, the uncertainty of the amplitude σ_{A_λ}, the phase ϕ, and the uncertainty of the phase σ_ϕ.

- The following values and options for the stellar models must be indicated: ranges for T_{eff} and $\log g$, stellar mass, metallicity, micro turbulence, source of the atmosphere grid, and the source of the non-adiabatic observables.

- The evolutionary stellar model grid for the indicated mass is searched for models that lie in the given range of T_{eff} and $\log g$.

- For each found model, the atmospheric parameters $\alpha_{T,\lambda}$, $\alpha_{g,\lambda}$, $\beta_{T,\lambda}$, $\beta_{g,\lambda}$, and $b_{\lambda,\ell}$ are determined by bi-linear interpolation in the grid of the indicated filter set, metallicity, and micro turbulence.

- The program searches in the lists of the found non-adiabatic pulsation models for different ℓ-values. For each ℓ, the frequency that is closest to the observed value is searched for. The values of the real and complex non-adiabatic parameters, f_R and f_I, respectively, are taken from this frequency value.

- The theoretical amplitudes and phases are computed from Eq. (31) for each selected filter.

- The amplitude ratios and phase differences are computed with respect to a selected filter (ideally the one with the largest observed amplitude).

FAMIAS creates an error message if no atmospheric or evolutionary models have been found in the grids for the indicated parameters.

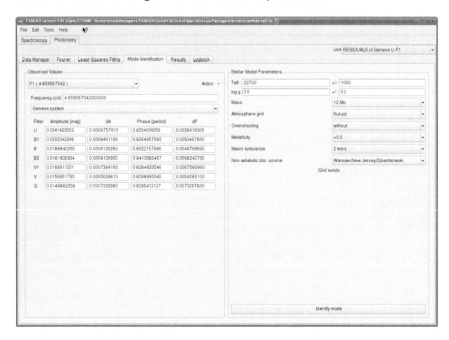

Figure 23: Screenshot of the *Mode Identification Tab*.

5.4.3 Observed values

This box contains the frequencies, amplitudes, and phases of the observed pulsation frequencies in different photometric filters. These values can be imported from the *Least-Squares Fitting Tab* or entered manually.

- **Frequency selection**
  Different frequencies can be selected with this combo box. Each item in the combo box is related to a frequency value and its amplitudes and phases in different filters. Frequencies can be added by importing from

the *Least-Squares Fitting Tab* or by selecting in the *Action menu* the option *Add frequency*.

- **Action menu**
  This menu allows to add or remove frequency sets.

- **Frequency value**
  Frequency value in d^{-1}.

- **Filter system**
  Select the filter system for which the mode identification should be carried out. Three different systems are available: Johnson/Cousins $UBVRI$, Strömgren $uvby$, and Geneva.

- **Table of amplitudes and phases**
  This table contains for each filter the observed values of A_λ and σ_{A_λ} in mmag, and of ϕ, and σ_ϕ in units of 2π. You do not have to fill out all fields. Empty fields (or $= 0$) are not used for the computation of the amplitude ratios and phase differences. Theoretical values are anyway computed for all filters.

5.4.4 Stellar model parameters

- **Teff**
  Observational value of the effective temperature in Kelvin and its uncertainty.

- **log g**
  Observational value of the logarithm of the gravity in c.g.s and its uncertainty.

- **Mass**
  Stellar mass in solar units. The available values depend on the selected non-adiabatic model source. You can only obtain a mode identification for one selected mass-value at a time.

- **Atmosphere grid**
  Model source of the grid of the atmospheric parameters $\alpha_{T,\lambda}$, $\alpha_{g,\lambda}$, $\beta_{T,\lambda}$, $\beta_{g,\lambda}$, and b_λ^ℓ.

- **Overshooting**
  This box indicates if models with core overshooting should be taken into account.

- **Metallicity**
  Stellar metallicity value $[m/H]$. The available range depends on the selected non-adiabatic model source.

- **Micro turbulence**
  Micro turbulence value of the stellar atmosphere models.

- **Non-adiabatic obs. source**
  Select here the source for the grid of non-adiabatic observables.

- **Identify mode**
  Start the mode identification. FAMIAS computes the observed amplitude ratios and phase differences as well as the corresponding values for all found pulsation models. The results are written to the *Results Tab*.

5.5 Results

This module contains the results of the photometric mode identifications. It gives the observed and theoretical values of the amplitude ratio and phase difference in different filters in a text window as well as in diagrams.

5.5.1 List of Calculations

Each time a mode identification is carried out, its results are saved as a new data set in this list. Click on an item to display the results in the field *Mode identification* and the corresponding diagrams in the field *Mode identification plots*.

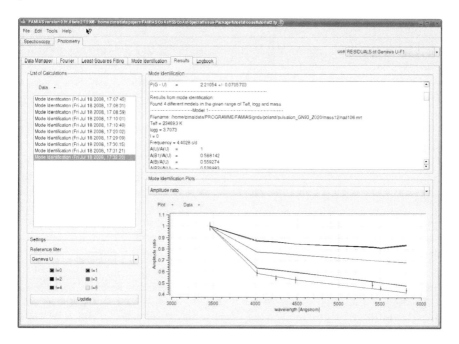

Figure 24: Screenshot of the *Results Tab*.

5.5.2 Settings

This box can be used to set the reference filter and to set which ℓ-values should be displayed in the plot window.

- **Reference filter**
  This is the reference filter r for the amplitude ratio and phase difference
  with respect to filter x. The amplitude ratios are computed as A_x/A_r.
  The phase difference is calculated as $\phi_x - \phi_r$. Exceptions are the mode
  identification plots, where the phase difference is plotted against the amplitude ratio. There, the indices r and x are exchanged.

- **Box of ℓ-values**
  You can select here which ℓ-values should be displayed in the plot window.

- **Update**
  This updates the *Mode identification box* and plot window with the
  current settings.

5.5.3 Mode Identification Report

This field displays the main information about the observed and theoretical
parameters for the obtained mode identification. It lists the input values and
settings for the models, the observed amplitude ratios and phase differences,
and for each pulsation model that matches the search criteria, its degree and
corresponding amplitude ratio and phase differences.

Amplitude ratios in the filters x and y are denoted as $A(x)/A(y)$. Phase
differences are indicated as $P(x - y)$ and in units of degrees. For the observed
values, the 1σ standard deviation is indicated.

5.5.4 Mode Identification Plots

Three kinds of plots, that can be selected via the combo box above, are available
in this field. They are described in detail below. In each plot, the observed
values are displayed as black crosses with error bars. The theoretical values are
displayed as lines. Each colour represents another value of the degree ℓ and
coincides with the colour-scheme in the field *Settings*. Generally, more than
one theoretical pulsation model is found that matches the input criteria (e.g.,
T_{eff} and $\log g$). All these models are displayed as apart lines in the plots (as
well as listed in the field *Mode Identification*).

- **Amplitude ratio**
  This plot displays the amplitude ratio $\frac{A_x}{A_r}$ relative to the reference filter r
  (see *Settings*) as a function of the central wavelength of the corresponding
  filter x. The values of each fitting theoretical model are drawn as apart
  lines. Plots of this kind are suited to identify modes in SPB or β Cep
  stars, i.e., stars where the amplitude ratio depends strongly on the degree
  of the mode.

- **Phase difference**
  This plot displays the phase difference $P(x - r)$ relative to the reference filter r (see *Settings*) as a function of the central wavelength of the corresponding filter x. Values of different fitting theoretical models are plotted as apart lines.

- **Phase diff / Ampl. ratio**
  These plots display the amplitude ratio $\frac{A_r}{A_x}$ as a function of the phase difference $P(r - x)$ for each filter x of the selected filter system relative to the reference filter r. Each plot displays the results for another combination of r and x. In these plots, the lines of a certain colour represent the range of all found models that fulfil the search criteria.

5.6 Logbook

The logbook shows the list of actions that have been performed with the photometric set of tools of FAMIAS and corresponding information. Each time an operation is carried out in FAMIAS, a new log-entry is written to the *List of actions*. Clicking on an entry of this list shows the corresponding information in the text box.

Entries of the *List of actions* can be renamed or deleted by using the menu *Data*. The text box can be modified and saved in FAMIAS by clicking on the button *Save*.

5.7   Tutorial: Photometric mode identification

This tutorial demonstrates the use of FAMIAS for the photometric mode identification based on multi-colour light curves.

5.7.1   Importing and preparing data

1. Select the *Photometry* page and click on *File → Import Light curve.*

2. Select one or several files that contain the photometric data. A data file must be in ASCII format and consist of at least two columns, separated by a space or tabulator. Columns of observation time in d^{-1} and magnitude are required. An additional column listing the weights of the measurements is optional. Once you have selected your files, click on *Open.*

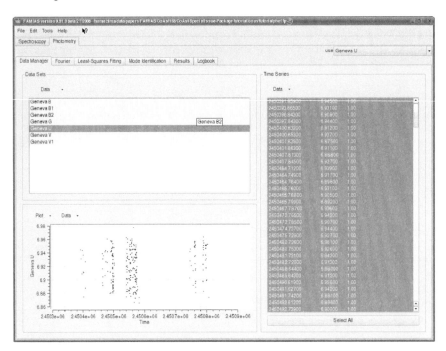

Figure 25: *Data manager* after importing Geneva light curves of an SPB star.

3. For each file that you import, an import-dialogue will open. Select the columns that you want to import and specify the photometric passband of

the observations. For a more detailed description of the import-dialogue see Section 2.1. The successfully imported data sets will be listed in the *Data Manager*.

4. You can use the tools in the *Data Manager* to edit your data (delete data points, change weights, etc.). See Section 5.1.2 for further details.

5.7.2 Searching for periodicities

The search for periodicities should be carried out in the following iterative schematic way:

1. Compute a Fourier spectrum in a frequency range where you expect pulsation.

2. Compute the significance level at the frequency having the highest amplitude and include this frequency in the least-squares fitting if it is significant.

3. Compute a multi-periodic least-squares fit of the original data with all detected frequencies. In case that no unique frequency solution exists due to aliasing, compute least-squares fits for different possible frequencies. The solution resulting in the lowest residuals should be regarded as best solution.

4. Exclude frequencies from the fit that do not have a SNR above 4 (3.5 for harmonics and combination terms).

5. Pre-whiten the data with all significant frequencies.

6. Continue with the first point using the pre-whitened data until no significant frequency can be found.

In FAMIAS follow the following procedure:

1. **Select data set**
   In the *Data Manager* or the *use* combo box (top right), select the data set you want to analyse.

2. **Calculate Fourier spectrum**
   Switch to the *Fourier Tab*. Select a reasonable frequency range and click on *Settings → Calculate Fourier* to compute a Fourier spectrum. A dialogue box will pop up and ask you, if you would like to include the highest frequency peak in the *Least-Squares Fitting Tab*. Before doing so, it is a good idea to check for the statistical significance of this peak.

3. **Compute significance level**

Mark the check box *Settings → Compute significance level*. The frequency value of the highest peak is automatically written to the corresponding text field. Modify this value, if you are interested in the significance of another frequency peak. Click on *Calculate Fourier* to compute another Fourier spectrum. The significance level will be shown in the plot as a red line. If the examined frequency peak is significant, include it in the *List of Frequencies* of the *Least-Squares Fitting Tab*.

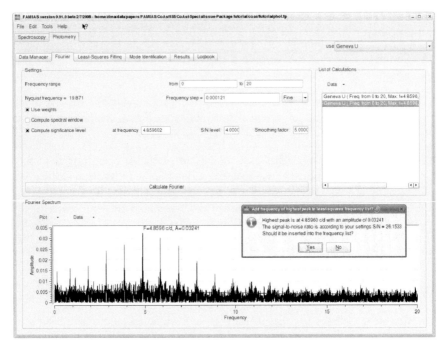

Figure 26: Fourier spectrum of the Geneva U-band.

4. **Compute least-squares fit**

Switch to the *Least-Squares Fitting Tab* and mark the check boxes of all frequencies in the *List of Frequencies* that you want to include in the fit. Also mark the check box *Settings → Compute signal-to-noise ratio* to determine the statistical significance of the selected frequency peaks.

Click on *Calculate Amplitude + Phase* to compute a least-squares fit to the data by improving amplitude and phase values. Click on *Calculate All* to compute a fit by improving frequency, amplitude, and phase. The *List of Frequencies* box shows the results of the fit. The frequency is displayed in inverse units of the input time and the phase is indicated in units of the period.

5. **Pre-whiten light curve**
   To pre-whiten the light curve with the obtained fit, mark the check box *Pre-whiten data* and compute another fit. The pre-whitened light curve is written to the *Data Manager Tab* as a new data set.

6. **Compute Fourier spectrum of residuals**
   Select the pre-whitened light curve and compute a Fourier spectrum thereof to search for further frequencies. If you want to compute another least-squares fit with additionally found frequencies, you must select the original light curve.

5.7.3   Mode identification

The photometric mode identification as it is implemented in FAMIAS uses the method of amplitude ratios and phase differences in different photometric passbands. You therefore have to provide for each pulsation frequency that should be identified its amplitude and phase for different filters. These values can be determined with a multi-periodic least-squares fit of sinusoids to the data under the assumption that the variations are sinusoidal. To be able to compare the phase values determined for the different filters, the same frequency values have to be taken into account in the least-squares fits for all filters. Important note: the frequency has to be in units of d^{-1}.

1. **Insert frequencies**
   You can either enter input the observed values of frequency, amplitude, and phase manually or copy directly the results from the least-squares fit.

   • **Manual input**
     Switch to the *Mode Identification Tab* (see Figure 27). In the field *Observed Values*, select the option *User input* in the top combo box. You can input the frequency value in the field *Frequency*. Select the filter system of your observations and input the observed amplitude and phase in the corresponding fields. To add values for an additional frequency, use the function *Action → Add frequency*.

   • **Automatic input**
     To use this function, you need to import light curves from different filters and assign to each light curve the correct filter name (during import or in the *Data Manager*). Mark the frequencies in the *Least-Squares Fitting Tab → List of Frequencies* that have significant

amplitude in all filters that you want to use for the mode identification. Press *Copy values to MI* to compute a least-squares fit and copy the frequency, amplitude, and phase to the *Mode Identification Tab*. In this least-squares fit, the frequency is kept constant whereas amplitude and phase are improved.

Repeat this procedure for the light curves taken in other filters without modifying the frequency values or the number of marked frequencies.

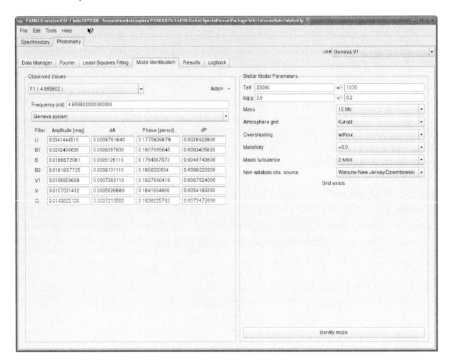

Figure 27: Mode Identification Tab of FAMIAS. The observed amplitude and phase are listed in the left field, whereas the options for the stellar models can be set in the right field.

For each imported frequency, a new item is added in the top combo box of the field *Mode Identification Tab → Observed Values*.

To carry out a mode identification, it is only obligatory to provide the observed frequency value. It is not necessary to input observed amplitude and phase values. The theoretical amplitude ratios and phase differences are in any case always computed for all filters of the selected filter system.

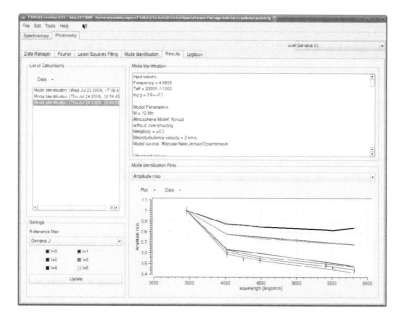

Figure 28: Results of the tutorial data. The best identification is achieved for $\ell = 0$ (red lines).

2. Set stellar model parameters

In the *Mode Identification Tab*, the parameters of the stellar models and the source of the non-adiabatic observables that should be used for the mode identification have to be set. See Section 5.4.4 for detailed information about these parameters.

3. Start mode identification

Start the mode identification by pressing the button *Identify mode*. The results will be written to the *Results Tab*.

4. Interpretation of results

The *Results Tab* displays the results of the mode identification in text form and in several plots. The text field *Mode Identification* lists the input parameters and the observed and theoretical amplitude ratios and phase differences. Its contents are described in detail in Section 5.5.

As can be seen in Figure 28, the observed amplitude ratios are most consistent with theoretical models that have a degree $\ell = 0$.

Bibliography

Aerts, C., de Pauw, & M., Waelkens, C. 1992, A&A, 266, 294

Aerts, C., & Waelkens, C. 1993, A&A, 273, 135

Aerts, C. & Eyer, L. 2000, in "Delta Scuti and Related Stars",
    eds. Breger, M., Montgomery, M. H., ASPC, 210, 113

Balona, L. A., & Stobie, R. S. 1979, MNRAS, 189, 649

Balona, L. A. 1986a, MNRAS, 219, 111

Balona, L. A. 1986b, MNRAS, 220, 647

Balona, L. A. 1987, MNRAS, 224, 41

Balona, L. A. 2000, in "Delta Scuti and Related Stars",
    eds. Breger, M., Montgomery, M. H., ASPC, 210, 170

Briquet., M., & Aerts., C. 2003, A&A, 398, 687

Claret, A. 2000, A&A, 363, 1081

Cuiger, H., Dziembowski, W. A., & Pamyatnykh, A. A. 1994, A&A, 291, 143

Daszyńska-Daszkiewicz, J., Dziembowski, W. A., Pamyatnykh, A. A.,
    & Goupil, M.-J. 2002, A&A, 392, 151

Daszyńska-Daszkiewicz, J., Dziembowski, W. A., & Pamyatnykh,
    A. A.2003, A&A, 407, 999

Dziembowski, W. A. 1971, AcA, 21, 289

Dziembowski, W. A. 1977, AcA, 27, 203

Dupret, M.-A., De Ridder, J., De Cat, et al. 2003, A&A, 398, 677

Lenz, P., & Breger, M. 2005, CoAst, 146, 53

Mantegazza, L. 2000, in "Delta Scuti and Related Stars",
    eds. Breger, M., Montgomery, M. H., ASPC, 210, 138

Martens, L., & Smeyers, P. 1982, A&A, 106, 317

Michalewicz, Z. 1996, Genetic algorithms + data structures = evolution programs,
    Springer-Verlag Berlin

Montalban, J., & Dupret, M.-A. 2007, A&A, 470, 991

Paczyński, B. 1969, BAAS, 1, 256

Paczyński, B. 1970, AcA, 20, 47

Schrijvers, C., Telting, J. H., Aerts, et al. 1997, A&A, 121, 343

Schrijvers, C., & Telting, J. H. 1999, A&A, 342, 453

Ventura, P., D'Antona, F., & Mazzitelli, I. 2007, Ap&SS, 420

Watson, R. D. 1988, Ap&SS, 140, 255

Zima, W. 2006, A&A, 455, 227

Acknowledgements

WZ has been supported by the FP6 European Coordination Action HELAS and by the Research Council of the University of Leuven under grant GOA/2003/04. Thanks to Conny Aerts, Maryline Briquet, Duncan Wright, and Jagoda Daszyńska-Daszkiewicz for the detailed proofreading of the manuscript that led to a significant improvement of the manual. Many thanks to Maarten Desmet for the many good ideas and to the asteroseismology team of the IvS for their critical testing of FAMIAS and their constructive comments and bug reports.

Comm. in Asteroseismology
Vol. 155, 2008

Index

Add white noise, 37

Break-up velocity, 26

Central wavelength, 55
Change assigned filter, 94
Change dispersion scale, 30
Clear tabs, 26
Combine data sets, 30, 94
Compute Fourier spectrum
  Light curve, 97
  Moments, 41
  Pixel-by-pixel, 41
Convert dispersion units, 36

Edit menu, 26
Equatorial rotation velocity, 26
Export data to ASCII-file
  2D-Fourier spectrum (pixel-by-pixel), 42
  Fourier spectrum (spectroscopy), 43
  Frequencies from least-squares fit, 48
  Least-squares fit (pixel-by-pixel), 49
Export spectrum, 30, 93
Extract spectral line, 36

File menu, 23
Fourier parameter fit method, 59
Frequency step, 40

Help menu, 26
Horizontal-to-vertical amplitude ratio, 26

Import light curve, 25
Import set of spectra, 23
Interpolate dispersion, 35

Limb darkening coefficient, 54

Mean spectrum, 31
Median spectrum, 31
Model grids, 104
Moment method, 63
Moments of line profile, 33

New project, 23
Non-adiabatic parameter f, 53
Nyquist frequency, 40

Open project, 23

Photometry
  Data manager, 93
  Data sets box, 93
  Fourier analysis, 96
  Least-squares fitting, 99
  Logbook, 111
  Mode identification, 102
  Plot window, 95
  Results, 109
  Time series box, 94
  Tutorial, 112
Photometry module, 93
Plot window, 28

Recent projects, 23
Requirements, 22
  Photometric data, 22
  Spectroscopic data, 22

Save project, 23

Save project as, 23
Shift dispersion scale, 37
Sigma clipping, 32, 37
Signal-to-noise level
    Fourier, 41
    Least-squares fit, 46
Signal-to-noise ratio, 32
Spectral window, 40
Spectroscopy
    Data manager, 29
    Data sets box, 30
    Fourier analysis, 39
    Least-squares fitting, 44
    Line profile synthesis, 50
    Logbook, 75
    Mode identification, 58

Plot window, 38
Results, 73
Spectrum box, 37
Time series box, 30
Tutorial, 76
Spectroscopy module, 29
Standard deviation spectrum, 31
Subtract mean spectrum, 37

Tools menu, 26
Tutorial
    Photometric mode identifica-
        tion, 112
    Spectroscopic mode identifica-
        tion, 76